Multilingualism in Mathematics Classrooms

BILINGUAL EDUCATION AND BILINGUALISM
Series Editors: Nancy H. Hornberger, *University of Pennsylvania, USA*, and Colin Baker, *Bangor University, Wales, UK*

Bilingual Education and Bilingualism is an international, multidisciplinary series publishing research on the philosophy, politics, policy, provision and practice of language planning, global English, indigenous and minority language education, multilingualism, multiculturalism, biliteracy, bilingualism and bilingual education. The series aims to mirror current debates and discussions.

Full details of all the books in this series and of all our other publications can be found on http://www.multilingual-matters.com, or by writing to Multilingual Matters, St Nicholas House, 31–34 High Street, Bristol BS1 2AW, UK.

BILINGUAL EDUCATION AND BILINGUALISM
Series Editors: Nancy H. Hornberger and Colin Baker

Multilingualism in Mathematics Classrooms
Global Perspectives

Edited by
Richard Barwell

MULTILINGUAL MATTERS
Bristol • Buffalo • Toronto

Library of Congress Cataloging in Publication Data
A catalog record for this book is available from the Library of Congress.
Multilingualism in Mathematics Classrooms: Global Perspectives/
Edited by Richard Barwell.
Bilingual Education & Bilingualism: 73
Includes bibliographical references and index.
1. Mathematics--Study and teaching. 2. Education, Bilingual. 3. Multilingualism.
4. Language and education. I. Barwell, Richard, 1969-
QA13.M87 2009
372.7'044–dc22 2009026146

British Library Cataloguing in Publication Data
A catalogue entry for this book is available from the British Library.

ISBN-13: 978-1-84769-205-4 (hbk)
ISBN-13: 978-1-84769-204-7 (pbk)

Multilingual Matters
UK: St Nicholas House, 31–34 High Street, Bristol BS1 2AW, UK.
USA: UTP, 2250 Military Road, Tonawanda, NY 14150, USA.
Canada: UTP, 5201 Dufferin Street, North York, Ontario M3H 5T8, Canada.

Copyright © 2009 Richard Barwell and the authors of individual chapters.

All rights reserved. No part of this work may be reproduced in any form or by any means without permission in writing from the publisher.

The policy of Multilingual Matters/Channel View Publications is to use papers that are natural, renewable and recyclable products, made from wood grown in sustainable forests. In the manufacturing process of our books, and to further support our policy, preference is given to printers that have FSC and PEFC Chain of Custody certification. The FSC and/or PEFC logos will appear on those books where full certification has been granted to the printer concerned.

Typeset by Techset Composition Ltd., Salisbury, UK.
Printed and bound in Great Britain by the Cromwell Press Group.

Contents

Contributors vii
List of Tables and Figures xi

1 Multilingualism in Mathematics Classrooms:
  An Introductory Discussion 1
  Richard Barwell

2 Mapping the Mathematical Langscape 14
  Frank Monaghan

3 Somali Mathematics Terminology: A Community Exploration
  of Mathematics and Culture 32
  Susan Staats

4 Politics and Practice of Learning Mathematics in Multilingual
  Classrooms: Lessons from Pakistan 47
  Anjum Halai

5 Mathematical Word Problems and Bilingual Learners
  in England 63
  Richard Barwell

6 How Language and Graphs Support Conversation in
  a Bilingual Mathematics Classroom 78
  Judit Moschkovich

7 Reflections on a Medium of Instruction Policy for
  Mathematics in Malta 97
  Marie Therese Farrugia

8 Bilingual Mathematics Classrooms in Wales 113
  Dylan V. Jones

9 Bilingual Latino Students, Writing and Mathematics: A Case Study of Successful Teaching and Learning 128
   Kathryn B. Chval and Lena Licón Khisty

10 Mathematics Teaching in Australian Multilingual Classrooms: Developing an Approach to the Use of Classroom Languages 145
   Philip C. Clarkson

11 Summing Up: Teaching and Learning Mathematics in a Multilingual World 161
   Richard Barwell

References 169

Index 181

Contributors

Richard Barwell is an Associate Professor at the Faculty of Education, University of Ottawa, Canada. His research is located in the intersection of applied linguistics and mathematics education, with a particular focus on multilingualism/bilingualism in the teaching and learning of mathematics. His research interests include mathematics classroom discourse, mathematics learning in multilingual settings and the relationship between learning language and learning curriculum content and his work has been published in peer-reviewed journals in applied linguistics, mathematics education and general education. Prior to his academic career, Dr Barwell taught mathematics in the UK and in Pakistan, where his interest in language and mathematics first arose.

Kathryn B. Chval is an Assistant Professor and Co-Director of the Missouri Center for Mathematics and Science Teacher Education at the University of Missouri, USA. Dr Chval is also a Co-Principal Investigator for the *Center for the Study of Mathematics Curriculum* and the *Researching Science and Mathematics Teacher Learning in Alternative Certification Models Project*, which are both funded by the National Science Foundation. Dr Chval's research interests are related to critical issues that impact the teaching and learning of mathematics including effective models and support structures for the teacher professional continuum, effective elementary teaching of underserved populations, especially Latinos and other English language learners and curriculum standards and policies.

Philip C. Clarkson has been at the Australian Catholic University since 1985, following nearly five years as Director of a Research Centre at the Papua New Guinea University of Technology. He has served as President of the Mathematics Education Research Group of Australasia and was Foundation Editor of the *Mathematics Education Research Journal*. Research interests include evaluation of schools, education systems and research programmes, as well as mathematics education. Professor Clarkson began his professional life as a secondary school teacher where he encountered immigrant students who spoke only Greek. This experience led to his interest in the interaction of language and mathematics learning.

Marie Therese Farrugia is a Lecturer in Mathematics Education at the University of Malta, Malta, where she contributes to the four-year Initial Teacher Education programme for primary teachers and to a post-graduate degree in Mathematics Education. Her main research interest is in the relationship between language and mathematics, including medium of instruction issues, the teaching and learning of meanings for mathematical expressions and language as a semiotic system. Dr Farrugia's interest in language finds its roots in her national linguistic landscape, where both Maltese and English are used in many social situations including mathematics classrooms. Dr Farrugia herself uses either or both languages according to the context at hand.

Anjum Halai is an Associate Professor and Head of the research and policy studies unit at the Aga Khan University Institute for Educational Development in Karachi, Pakistan. She has a wide experience of teaching in schools and in teacher education at graduate and post-graduate levels. Dr Halai was the Founding Chairperson of the Mathematics Association in Pakistan. Her research interests include social justice in mathematics teacher development and in students' learning of mathematics. Her doctoral study (Oxford University, UK) focused on the role of social interactions in students' learning of mathematics in multilingual classrooms in Pakistan. She speaks Gujrati, Urdu and English and understands Punjabi and Seraiki.

Dylan Vaughan Jones is Head of the School of Initial Teacher Education and Training at Trinity University College, Carmarthen, Wales, UK. His research interests lie in the learning, teaching and assessment of mathematics within bilingual contexts, and he has published a number of articles in this field. Previously, Dr Jones' has taught mathematics in a Welsh bilingual secondary school, worked as a secondary school inspector, and directed the post-graduate certificate of education programme at the University of Wales Aberystwyth. Dylan, his wife Nia and three teenage boys, Tomos, Ifan and Gerallt, are all first-language Welsh speakers and live near Aberystwyth in mid-Wales.

Lena Licón Khisty is Associate Professor in Curriculum and Instruction at the University of Illinois at Chicago (UIC), USA, where she teaches graduate and undergraduate courses in both bilingual/ESL education and mathematics education. Her research interests include interaction and discourse processes that impact Latina/o students in mathematics, especially from the perspectives of sociohistorical activity and critical race theories. She has had several federally funded projects for teacher development related to implementing integrated first and second language development and mathematics. She is currently a Principal Investigator for the *Center for the Mathematics Education of Latinos*, a consortium of four universities funded by the US National Science Foundation.

Frank Monaghan is a Senior Lecturer in Education and Language Studies at The Open University in London and Vice Chair of NALDIC, the United Kingdom's professional association for those working with learners of English as an additional language. Before joining the Open University, he spent 20 years teaching English as an additional language in one of London's largest and most linguistically diverse secondary schools, working mainly in mathematics. His PhD was inspired by this experience. His current research interest is in dialogic teaching and learning in multilingual mathematics classrooms. Frank first experienced being a learner through an additional language himself when spending a term as an exchange student in Germany, aged 16.

Judit Moschkovich is Associate Professor of Mathematics Education in the Education Department at the University of California, Santa Cruz, USA. Her research focuses on mathematical thinking and learning and Vygotskian approaches to learning. She is originally from Argentina, her first language is Spanish, and she moved to the United States in high school. Dr Moschkovich has worked as a teacher and researcher with students of Mexican, Puerto Rican and Central American origin living in the United States. She currently conducts research in secondary mathematics classrooms, is a Principal Investigator for the Center for the Mathematics Education of Latinos/as (CEMELA), funded by the US National Science Foundation.

Susan Staats is a mathematics educator and a cultural anthropologist at the University of Minnesota, USA. During her field research in indigenous Guyanese communities in South America, Dr Staats studied the Kapon language and analysed communicative ideologies in religious discourse. At the University of Minnesota, Dr Staats teaches algebra and interdisciplinary seminars to first-year students. She studies students' use of language as a means to construct shared contexts of knowledge in inquiry-based algebra lessons and in interdisciplinary mathematics applications.

List of Tables and Figures

Tables

2.1	Ethnic backgrounds of students of a London secondary school	16
2.2	Frequency table of words from the field of 'area, shape and volume'	22
2.3	Reflections of broader culture in SMILE materials	30
6.1	Multiple meanings for 'went by'	92
8.1	A possible approach to a concurrent bilingual lesson	119
9.1	Frequency of word use by Sara and her students	138
9.2	Comparison of median grade equivalent on the Iowa Test of Basic Skills (ITBS) mathematics total for Sara's students with other fifth grade students in their school, district and nation	143
10.1	Mapping languages used in Australian urban classrooms	154

Figures

2.1	Screenshot of Cobuild corpus illustrating use of 'diagonal' in general use	21
2.2	Use of 'diagonal' in its various forms in SMILE materials	24
2.3	Smile Card 1112 – rotation	26
2.4	Screenshot of concordancer showing use of 'about' in SMILE mathematics materials	27
2.5	SMILE 1858: Number jigsaw originally developed by a 'new arrival' bilingual student	28
2.6	Pie-chart activity	29

4.1	The Rooh Afza problem	52
5.1	An arithmetic word problem (QCA, 1998)	63
6.1	Problem: From Atlantic City to Lewes (*Connected Mathematics*, Lappan et al., 1998)	82
6.2	Carlos' graph	84
6.3	David's graph	84
6.4	'I went by twos' describing number of segments	89
6.5	'I went by twos' describing number of units	89
6.6	Carlos describes his scale as 'I went by fives'	90
6.7	Carlos describes David's scale as 'He went by one'	90
6.8	Teacher describes Carlos' scale as 'You went by two and a halves'	91
8.1	The formation of teaching groups in a traditional bilingual secondary school	117
9.1	Violetta's first draft of the missing leg problem	130
9.2	Violetta's first draft of the semicircle problem (page 1)	131
9.3	Violetta's first draft of the semicircle problem (page 2)	132
9.4	Javier's second draft	140
9.5	Matthew's third draft	142
10.1	Language use in mathematics learning	148
10.2	A modification to the language use in mathematics learning model	149
10.3	A model for language use in mathematics learning for multilingual students, with the overall flow of language downwards	150

Chapter 1
Multilingualism in Mathematics Classrooms: An Introductory Discussion

RICHARD BARWELL

Farida is a student in a medium-sized urban primary school in the United Kingdom. She has attended the school since she joined the nursery class and is now in Year 5 (9–10 years old). Her family, which is from Pakistan, lives near to the school and her parents work for long hours in the shop they run. At home Farida speaks Punjabi, Urdu and English at different times. When I first met Farida, she told me a bit about the mathematics she had recently been working on, including the following ('she' refers to her teacher, Miss T):

> oh yeah, circle, and shapes and she talks about three Ds and two Ds like, but one face and hexadas she like says six sides yeah and, and, Miss T like, choose hepsadas, you have six sides, yeah? and pentagon has eight sides, um, and ummm, ummm ...

What mathematics can you see in what Farida says? What language issues might arise? If you were her teacher, what mathematics and what language would you want to work on with her? When she says 'pentagon has eight sides' is that a language issue? A mathematics issue? Both? Or just a slip of the tongue?! More generally, does multilingualism have any effect on the attainment of students like Farida in mathematics? How does multilingualism affect how such students participate in mathematics lessons? What role do students' different languages play in their learning of mathematics? What can we do as teachers' to support students like Farida? How can we balance teaching mathematics with teaching Farida the language of mathematics?

This book explores many of these questions, drawing on research and practice in a variety of different multilingual mathematics contexts. The book has two related aims. One is to give a sense of the diversity of what multilingual mathematics classrooms can be like. The case of Farida gives some sense of what multilingualism in one mathematics classroom in the United Kingdom might involve. The second aim of the book is to explore issues arising in these particular contexts in such a way that this exploration informs practice across contexts. This aim is not about providing simplistic generalisations or recipes for teaching. Language, learning and society are too complex for that. Rather, by discussing issues that arise in one context, readers will, I hope, encounter new perspectives and new ways of thinking about their own.

Later in this introduction, I introduce the nine contributions that make up the main part of the book. Before that, however, I want to set the scene. First, I provide an overview of research findings on a key question that is not significantly addressed in the chapters that follow: does multilingualism make any difference to students' attainment in mathematics. Second, I briefly highlight some of the key ideas that have emerged in previous classroom research concerned with multilingualism and the teaching and learning of mathematics, since these ideas have influenced much of the more recent work presented in this book. I cannot move on to address any of these points, however, without first clarifying the focus of the book.

What Makes a Mathematics Classroom Multilingual?

Farida's experiences illustrate the complexity of even defining what 'multilingualism in mathematics classrooms' might be about. In a time of migration, global mobility and concern for minority rights, multilingualism is ubiquitous, yet almost infinitely varied. From the linguistic diversity arising from immigration to cities like London or New York, to the multilingualism of a South African township, from the relocation of a doctor or business executive and their family to Milan or Adelaide or Bangalore to the millions of Chinese who have learned English without ever having left China, multilingualism has many faces. For this book, and in the context of mathematics classrooms, multilingualism simply refers to the presence of two or more languages (and so includes bilingualism). Such a presence may be overt or tacit. That is, mathematics classrooms are considered to be multilingual if two or more languages are used overtly in the conduct of classroom business. And mathematics classrooms are *also* considered to be multilingual if students *could* use two or more languages to do mathematics, even if this does not actually occur, as Farida experienced in the United Kingdom.

This definition of a multilingual classroom encompasses a wide range of situations. In the United Kingdom, Farida was considered to be a learner of English as an additional language (EAL), a label that privileges English and obscures the anonymous 'additional' languages. In some parts of the world, other terms, including English as a second language (ESL) or English language learner (ELL), would be used instead, although they have similar drawbacks. In such situations, multilingual students are expected to learn the language of schooling, occasionally in addition to some of their other languages, more usually with indifference to them. The definition also includes classrooms found in many parts of the world in which students routinely use a mixture of two or more languages both in school and in wider society, such as in bilingual education programmes in North America or in some schools in Wales, or as in highly multilingual societies like South Africa or India. In some of these situations, students are not necessarily learners of the official language of schooling; it is simply one of several languages in which they are proficient. Although the most common concern of teachers and researchers is with situations in which multilingual students are learners of the classroom language, it should not be overlooked that this need not be the case.

Multilingualism and Attainment in Mathematics

Perhaps the most fundamental question asked by teachers, policy makers and researchers alike is 'does it make any difference'? Does multilingualism have any effect on mathematical attainment? There have been few large-scale surveys of mathematics attainment that investigate multilingualism as a factor. This is, perhaps, because variations in language proficiency, language structure and background social, cultural and political conditions make it difficult to attribute any difference in performance in mathematics even partly to multilingual factors. It is difficult to say whether Farida's performance in mathematics is affected by her proficiency in English, the fact that she has spent time in two different countries, the fact that she is a member of what is regarded as a minority ethnic group in the United Kingdom, the educational or economic status of her parents or a host of other possible factors. Nevertheless, although far from conclusive, large-scale studies have led to the concern that, in many contexts, multilingual students tend to underachieve in mathematics (Cocking & Chipman, 1988; Hargreaves, 1997). In the United Kingdom, for example, Phillips and Birrell (1994) found that a sample of EAL students from South Asian backgrounds had lower mathematics attainment than monolingual students and that they made less progress in mathematics over a year. These findings were in contrast to the same EAL students' performance in literacy tests, where they did as well as their monolingual counterparts. In a much larger study conducted in South Africa, where there are 11 official languages, Howie (2002, 2003) compared the mathematics attainment and English proficiency of more than 9000 secondary school students using data collected for an international comparison of mathematics and science performance (called TIMSS-R). In South Africa, the vast majority of schools use English as the formal teaching language for subjects like mathematics, so English is the language for textbooks and examinations.

In most mathematics classrooms, however, several different languages can be heard (see, e.g. Setati & Adler, 2000). Howie found that higher scores on the English proficiency test were correlated with higher scores on the mathematics test and that proficiency in English was the most significant factor in explaining differences in students' mathematics scores.

Even where mathematics attainment appears to have some connection with proficiency in a second language like English, as in Howie's study, it is not clear whether differences are a result of linguistic, cultural, social or economic conditions or some combination of these and other factors (Secada, 1992). One complication, for example, is that the tests used to measure mathematics attainment in the research described above are written in English, so that students' proficiency in English may obscure their attainment in mathematics. A further issue in reporting the attainment of multilingual students is that these studies implicitly take the majority or dominant perspective when considering what counts as attainment and what counts as mathematics, downgrading other forms of mathematical achievement as less valuable. Farida may be able to do arithmetic in Punjabi, for example, but this will never be tested at her school in the United Kingdom. She may be familiar with mathematical practices from her parents' shop, but these practices may not be recognised or considered in her mathematics lessons. Often, the only attainment that counts is that measured by standardised tests written in the societally dominant language.

The impact of multilingualism on mathematics attainment is far from straightforward and the role played by language in a mathematics classroom is complex. A number of different researchers have attempted to reduce this complexity by focusing in more depth on the relationship between language proficiency and mathematical attainment. Proficiency is a logical factor to consider. Language is unlikely to be an issue for a student who is fluent in the classroom language, although there may be related issues of a cultural and social nature. For a student who is new to or learning the classroom language, on the other hand, there are clearly challenges in participating in and learning mathematics. Between these two positions lie the majority of EAL students in the United Kingdom, for example, or students joining immersion programmes, such as in many African or Asian contexts. In Farida's case, we can speculate that some aspects of learning mathematics may be problematic. Her attempts to say the word 'hexagon' suggest that, for her, mathematical discussion may be more convoluted and involve more attention to the language and so less attention to the mathematics than for her monolingual peers.

Researchers interested in language proficiency and mathematics attainment have been influenced by the work of Cummins (2000a, 2001), especially his 'threshold hypothesis'. The threshold hypothesis states that for multilingual students, having low levels of proficiency in all their languages is a cognitive disadvantage. Being highly proficient in two or more languages gives cognitive advantages to students, while proficiency in only one language offers neither advantage nor disadvantage. A number of studies have been based on the assumption that any cognitive advantage or disadvantage will show up in students' levels of mathematics achievement.

A key early study was conducted by Dawe (1983) in the United Kingdom in which he investigated the threshold hypothesis by testing groups of around 50 students aged 11–14 years from four different language backgrounds: Punjabi, Mirpuri, Jamaican Creole and Italian. As measures of linguistic proficiency, he used tests of English reading comprehension and tests of competence in their first language (L1). For mathematical performance, Dawe used a test of deductive reasoning or logical thinking set in English and a test of logical connectives, that is problems involving words like 'if ... then', 'either ... or' and 'but'. By comparing scores on the linguistic tests with those on the mathematical tests, Dawe did find evidence to support Cummins' thresholds, particularly the lower threshold. Students who did not score highly on either English or their L1 did not generally score highly in the mathematics tests. He also found some evidence to support the upper threshold. Interestingly, this effect was strongest in the Mirpuri group, despite the relatively low social status of Mirpuri as a dialect of Punjabi.

Since Dawe's (1983) work, a number of other studies have provided further evidence that linguistic proficiency is related to mathematical attainment, in line with Cummins' ideas. Clarkson, for example, has conducted several studies in Australia and Papua New Guinea involving multilingual students in upper primary school (e.g. Clarkson, 1992, 2007; Clarkson & Galbraith, 1992). His approach involved testing students' linguistic proficiency in English and in their L1[1] as well as on different aspects of mathematics, including mathematical word problems. He used the scores from the linguistic proficiency tests to divide the students into three groups. The low–low (LL) group consisted of students with low scores in two languages and the high–high (HH) group consisted of students with high scores in two languages. In line with Cummins (2000a, 2001), the third group consisted of students with a high score in one language only, which he called 'one dominant'. Clarkson (1992, 2007; Clarkson & Galbraith, 1992) then looked at the mathematics scores of the three groups of students. The LL groups recorded significantly lower scores than the other two groups on at least some aspects of mathematics in each study, supporting the idea of a lower threshold. Clarkson's research also provides evidence to support the upper threshold, although the link is less strong than that for the lower threshold.

Overall, Clarkson's (1992, 2007; Clarkson & Galbraith, 1992) and Dawe's (1983) work seems to show that students' proficiency in both (or all) of their languages does make a difference to their performance in some mathematical tasks. Students who are highly proficient in two or more languages are likely to do better than average, while strong proficiency in at least one language appears to be an important factor in ensuring that multilingual students match monolingual students in mathematics attainment. In the case of the latter, it is important to note that this proficiency does not have to be in the classroom language; proficiency in other languages can be just as valuable. In Farida's case, for example, one way of enhancing her level of attainment in mathematics might be to support the development of her proficiency in Urdu and Punjabi.

The research discussed above provides the backdrop to the work presented in this book. While research shows that multilingualism does have an effect on performance in mathematics, it offers less in the way of explanation. Clarkson (2007) has suggested that proficient bilingualism enhances students' meta-cognitive skills in mathematics, that is, it allows students to think more effectively about their mathematical thinking. This idea is supported by experimental research in psycholinguistics (for a review, see Moschkovich, 2007) that suggests, for example, that the advantages of multilingualism include an enhanced capacity to analyse problems and select useful information, while ignoring other less useful features (e.g. Bialystok, 1992, 1994). As Moschkovich (2007) argues, however, there is a tendency to reduce questions about the role of multilingualism in the teaching and learning of mathematics to questions about individual cognition. As I have already implied, social factors are also likely to be important. In particular, the work discussed above says little about what goes on in multilingual mathematics classrooms and about how multilingualism is implicated in the process of teaching and learning mathematics. This focus is one that concerns all of the contributors to this book. Before introducing their specific contributions, I will discuss some of the research that has previously been conducted on multilingualism in mathematics classrooms and that has informed much of the work in the chapters that follow.

Multilingualism and Learning and Teaching Mathematics

Classroom research that investigates multilingualism and the teaching and learning of mathematics has led to the identification of a number of tensions that arise when teaching mathematics in multilingual classrooms (see, in particular, Adler, 2001). These tensions appear to have relevance across a wide range of contexts. In this section, I highlight three of these tensions:

- Tension 1: between mathematics and language;
- Tension 2: between formal and informal language;
- Tension 3: between students' home languages and the official language of schooling.

During research in the United States, both Khisty (1995) and Moschkovich (1999a) explored the question of what teachers can do to facilitate bilingual student participation in mathematical discussions. In her ethnographic study of three Spanish–English bilingual mathematics classrooms, Khisty (1995) found differences in how teachers attended to the language of mathematics, particularly where potential ambiguities arise in *both* Spanish or English. Teachers who seemed to be more effective paid more attention to the language of mathematics as well as to the mathematics itself. Moshchkovich (1999a), meanwhile, studied a third-grade class of Spanish-speaking EAL students taught by a bilingual teacher. She noticed a key element in how the teacher managed the discussion was to ensure that the focus stayed on mathematics, particularly by listening to the mathematical content, however it was expressed. These studies seem to suggest that in some situations, it is important to pay careful attention to how students express their mathematical ideas, while in others, it is important to engage carefully with the mathematics itself. Moschkovich (1999a) is critical of some 'advice' for teachers, which, she argues, focuses excessively on mathematical vocabulary and the use of real or 'concrete' objects to support students' learning. It may be that engaging with students' mathematics is more productive than simply teaching students vocabulary. We can see the force of this position in the case of Farida. As we saw in the opening extract, she has acquired a great deal of reasonably accurate vocabulary, but is perhaps still working on how to use some of it!

For teachers, then, there is a difficult balance to be struck between attention to *mathematics* and attention to *language*, where the latter includes not only vocabulary but also broader aspects of language such as mathematical ways of talking, arguing and explaining. Striking this balance emerged as a key issue for teachers participating in research conducted by Adler (2001), a teacher educator in South Africa. Adler worked with six teachers from different school backgrounds in South Africa. Through this work, she developed what she calls a 'language of teaching dilemmas', which emerged for her teachers in their multilingual classroom environments. One of the teaching dilemmas elaborated by Adler, that of 'transparency', relates to the preceding discussion of mathematical language. In this context, transparency concerns the visibility of, or explicitness of attention to, mathematical language. This issue is raised by one of the teachers participating in the study, who reflects on an incident when a student was giving an explanation to the rest of the class:

> [she] put forward what she thinks is going on in relation to that issue and it is a question of even though her language is not clear is there understanding amongst the rest of the students and it seems like the rest of the students do understand even though she is using incorrect language. (Adler, 2001: 130)

Here, there is a tension for the teacher between teaching 'correct' formal mathematical language and her students learning mathematics. The teacher's incident could also illustrate a second dilemma articulated by Adler: whether or not to intervene. By intervening, the teacher could shift the attention of the students from the mathematics they are grappling with to the language used to explain that mathematics. Intervening could allow the student giving the explanation to do so using more formal mathematical language, but the resulting explanation will not necessarily be the explanation she would have given for herself. In some ways, intervention would disempower her student. On the other hand, offering a language for students can empower them to develop their thinking. We can see how these dilemmas may impinge on Farida's learning of mathematics. Consider, for example, the moment when she says 'hexadas she like says six sides yeah and, and, Miss T like, choose hepsadas, you have six sides, yeah? and pentagon has eight sides'. How would you respond? Do you correct her 'hexadas' and 'hepsadas' to the word 'hexagons'? Or do you focus on her assertion that 'pentagon has eight sides'. And if so, are you thinking that pentagons have five sides, not eight. Or that octagons have eight sides, not pentagons. Are you responding to mathematics here or to language? How else could you respond?

Mathematics classrooms are, of course, situated within wider sociolinguistic contexts and language use in classrooms is more than a simple instrument used to feed mathematics into students. Farida, for example, had also attended school in Pakistan where mathematics teaching was more formal and memory-based, with strict discipline enforced by the teachers (see, e.g. Emblen, 1988), and where the lessons involved the use of Urdu and Punjabi. The role of language and languages, both in mathematics and in wider society, is, in many ways, quite different in the two countries in which Farida attended school. The significance of these broader aspects of language use in the teaching and learning of mathematics is illustrated by a study conducted by Setati (2005a), also in South Africa. Setati examined the interaction in a primary school mathematics lesson taught by a teacher, Kuki, who used at least two languages in her teaching, English and Setswana. In the lesson, the class were working on an arithmetic word problem about the number of dogs at the Society for the Prevention of Cruelty to Animals (SPCA). Setati's analysis shows that the use of English was largely restricted to 'procedural discourse', in which mathematical procedures such as methods of arithmetic calculation were rehearsed. Setati argues that this procedural discourse combines with the dominant position of English in South African society to produce an authoritative view of mathematics as being concerned with algorithms and right answers. Setswana, meanwhile, was used as a marker of solidarity. The solidarity of Setswana was combined with the occurrence of 'conceptual discourse' in which students and teacher discussed issues such as what the problem was about and how it could be interpreted mathematically.

For Setati, these findings illustrate the political dimension of language use in multilingual mathematics classrooms. The teacher's decisions are not purely pedagogical; they also reflect a wider political context in which English is seen as a valuable language and the natural language of mathematics, while Setswana is not seen as a language for doing formal mathematics. Setati (2005b) points out that the political dimension of language use leads to what is often seen as a tension. On one side of this tension is a desire to learn English as a prestige language that allows access to higher education or well-paid jobs. On the other side is the use of students' home languages for teaching mathematics, through which they are more likely to develop a deeper conceptual understanding of mathematics.

The three tensions that have emerged from recent classroom research show the complexity inherent in multilingual mathematics classrooms. None of these tensions can be resolved; teachers live with them through their daily practice. How they play out may vary from teacher to teacher, from school to school and from place to place. But the tensions themselves have a more general relevance, allowing us to focus on particular aspects of our practice and develop strategies suited to our own particular circumstances and needs. The chapters in this book explore different aspects of these and other tensions, each in a particular context, but each with something of a wider value to say.

Global Perspectives on Multilingualism in Mathematics Classrooms

This book is informally organised into three parts, although there is considerable interconnection between all the chapters. In the first two chapters, there is an emphasis on understanding the nature of mathematical language and on the tools that can help to do this. The next three chapters place more of an accent on students' responses to mathematical tasks in multilingual classrooms, although all three then seek to draw broader conclusions for mathematics teaching. The next four chapters look at policy and practice in multilingual mathematics classrooms from a variety of perspectives, highlighting challenges and exploring possible ways of addressing these challenges. All of the contributions are derived from the authors' research as well as from their experiences as teachers, learners and, in most cases, multilinguals. Each of them seeks to offer practical responses to the tensions that arise in teaching mathematics in multilingual settings.

One of the tensions discussed above concerns the extent to which teachers pay attention to mathematical language. For this to be possible, we need a reasonably clear idea of what the language of school mathematics is like. The first two chapters propose different methods with which to find this out. In Chapter 2, *Frank Monaghan* explores the nature of mathematical English found in a highly multilingual school in London, United Kingdom. Descriptions of mathematical English date back several decades (see Halliday, 1978; Pimm, 1987). Monaghan argues that this earlier work would be enhanced by more systematic analysis of the regularities of mathematical English (or any other language for that matter) based on actual examples of language use. The tools for such an analysis have been developed to a high degree of sophistication in a branch of applied linguistics called corpus linguistics. Corpus linguistics works by collecting a large sample of texts or recordings and then analysing them using specialised 'concordancing' software. Monaghan clearly describes how this kind of analysis works and how it can be useful in multilingual mathematics classrooms. He ends his chapter by showing how the results of corpus analysis can inform practice, such as in the development of new resources for multilingual learners of mathematics.

Susan Staats (Chapter 3) works in Minnesota in the United States, in a context that is in some ways similar to Monaghan's. As in London, multilingual students are expected to learn English. In Staats's case, however, she has sought a deeper understanding of how mathematics is expressed in a language used by some of her students. That language is Somali. Her chapter focuses on the nature of mathematical vocabulary in Somali, exploring some of the metaphors through which everyday Somali words are extended into mathematics. These metaphors include the beautiful linking of the brackets used in algebraic notation with crescent moons. Staats goes on to consider what her findings suggest about the relationship between culture and mathematics. In addition, she gives an account of how she used anthropological methods to investigate the nature of Somali mathematics vocabulary, thereby providing a valuable model for others to follow.

The significance of what particular words mean in different languages becomes a key issue in *Anjum Halai*'s chapter, based in Karachi, Pakistan. In this context, children use at least two languages in their mathematics lessons: Urdu and English. Halai begins by exploring how students' interpretations of particular words across languages can become significant in their understanding of a mathematical problem. For example, their understanding of a problem about making a drink by diluting a concentrate depends on which Urdu word they use in their small-group discussions to refer to the English word 'stronger'. She goes on to show how other aspects of language are also implicated in this cross-linguistic meaning making, including verb tenses, underlining the point that mathematical language involves more than vocabulary. Halai links this apparently local tension between using Urdu and using English to learn mathematics to the wider politics of language in Pakistan.

Like Halai, in my own contribution (Chapter 5), I discuss how multilingual students interpret mathematical problems, this time in England. For such students, there is, perhaps, a tension between the rather artificial nature of the problems and the subtext that says that real-world thinking is not required. This tension can be particularly difficult to navigate for multilingual students. In my chapter, I summarise some of the ways in which one particular task appeared to support multilingual students in making sense of word problems. The task allowed students to draw on personal experience of the world, promoted explicit attention to and discussion of the form of word problems, and led to a productive interaction between language learning and mathematical thinking. I illustrate these ideas using a transcript of two students, one a new learner of English, working on the task to produce a viable word problem that the multilingual participant is then able to solve with little difficulty.

In Chapter 6, *Judit Moschkovich* develops the idea that a focus on mathematical meaning rather than language *per se* supports the mathematical thinking and learning of multilingual students. She discusses an episode in which two Spanish–English students in a school in the United States work on a graphing task. Moschkovich traces the various different interpretations concerning the scale on the axes of the graph that arise in the discussion between the two students and their teacher. For Moschkovich, the students' interpretations are resources used for making mathematical meaning. By working with these interpretations, rather than evaluating them, the teacher allows their understanding to develop without necessarily needing direct access to that understanding. This is a powerful model for any mathematics teacher who, for example, works with students who are learners of the classroom language.

In Chapter 7, *Marie Therese Farrugia* writes about mathematics teaching in Malta, where Maltese is the national language, but English is also widely used. Her starting point is the Maltese government's policy on the use of Maltese and English in schooling, specifically in relation to mathematics. The policy recommends the use of English as the language of mathematics teaching. Using examples from primary school mathematics classrooms, Farrugia identifies and explores a number of tensions within the official policy. She argues that the recommendation to use English is in tension with other principles, such as the benefits of cooperative group work, the need for an inclusive learning environment and the desirability of the strengthening of the Maltese language. Her chapter raises issues that are likely to arise in mathematics curriculum and policy in many other parts of the world.

Dylan V. Jones, in Chapter 8, also writes about a bilingual environment, this time in Wales, United Kingdom. Drawing on his experience as a teacher, teacher educator and researcher, he discusses a range of issues that arise from the varied nature of Welsh–English bilingualism found in

Welsh schools. These issues include the challenges of organising mathematics classes when there is a range of Welsh language proficiency present within a cohort of students, the nature of classroom talk and the challenges of bilingual mathematics assessment. The chapter illustrates particularly well the link between the daily business of teaching and learning mathematics in classrooms and the wider sociolinguistic context. Indeed, this work implies another tension: even as the use of Welsh to teach mathematics contributes to a broader revival in the Welsh language, it creates challenges in the classroom.

Kathryn B. Chval and *Lena Licón Khisty* (Chapter 9) provide a case study highlighting the effective use of students' writing. The teacher, who teaches a class of students from Latino backgrounds in the United States, uses several strategies to support the development of students' mathematical thinking and mathematical language through writing. In particular, she uses a drafting process commonly found in subject English teaching, but rarely used in mathematics. Through this approach, students come to clarify their ideas and, therefore, develop more effective ways of expressing these ideas. This writing strategy is linked with several others, both written and oral, that are unified by a concern with mathematical meaning. Chval and Khisty show how these strategies combine to successfully integrate second-language learning with mathematics learning.

In Chapter 10, *Philip C. Clarkson* draws on his extensive research in Australian mathematics classrooms, where multilingual students are expected to learn and use only English. In his recent research, Clarkson has discussed with multilingual students how they draw informally on their different languages as they work on mathematics. Some students, for example, reported using a home language to make sense of aspects of a problem. Clarkson goes on to set out a model designed to promote the structured use of multilingual students' different languages as they go through the process of learning mathematics while learning English. The model involves moving from students' informal mathematical talk in some or all of their languages to the development of the conventional academic language of mathematics, again in some or all of their languages. Clarkson concludes by discussing some starting points for implementing such a model.

Reading through all the chapters, it is noticeable how some issues or ideas recur in different situations. The book concludes, therefore, with a discussion (Chapter 11) of some of these common threads, including some considerations of the possible implications for mathematics teaching practice.

I began this introductory chapter with a short vignette about Farida, one multilingual student in a primary school in the United Kingdom, talking about some of the mathematics she had recently been working on. This example is just one instance of multilingualism in a mathematics classroom, one instance of how several languages, overtly or tacitly, may be part of students' learning of mathematics. The chapters that follow offer nine different examples, with discussion and analysis of the issues arising in each one. It is said that travel broadens the mind. This book is an opportunity to travel and experience nine different multilingual mathematics classrooms.

Note
1. In Papua New Guinea, the L1 was designated as pidgin, which we must assume was the students' second language.

Chapter 2
Mapping the Mathematical Langscape

FRANK MONAGHAN

Learning mathematics also involves learning the language of mathematics. The two do not exist in isolation, but are part of a living, dynamic intellectual space, experienced and jointly constructed by real people engaging with mathematics and each other through the medium of language. Just as we negotiate our way through new landscapes by means of travel guides, maps and helpful passers-by, so too do we chart our course through what I term the *mathematical langscape*, the combination of mathematical meanings and the resources for communicating these meanings that make up the mathematics curriculum. We chart this course largely through our interaction with the verbal and visual language of mathematics and with those who teach and learn it alongside us. These meaning-making resources are referred to by linguists as the *register* of mathematics, that is to say, the way language, symbols and other visual representations are used to create the particular meanings found in mathematics as opposed to, say, science or history. In this chapter, I will focus on some specific aspects of the *mathematical langscape*, seen through the lens of my experiences of particular children and multilingual classrooms in London, United Kingdom, in the hope of illuminating some of the peculiarities and generalities of language in the multilingual mathematics classroom.

The linguist Michael Halliday has argued:

> The core of the difficulty in the mathematics classroom is that the teacher often understands and takes for granted the whole register of mathematics, and thinks only of the mathematical aspects of these items, whereas for the learner they may also be unfamiliar language – they are 'peculiar' English. It is therefore desirable that the mathematics teacher should be aware of the register of mathematics as a sub-set of English... To this end, mathematics educators and... English language teachers should collaborate in the production of guidelines, illustrative descriptions and teaching materials concerned with this problem. (UNESCO, 1975: 121)

Halliday's suggestion is, perhaps, even more desirable in the case of multilingual mathematics classrooms, given that some 50% of the world's population now regularly use more than one language in their daily lives (Franceschini, 1998), and that the language used in the classroom is not necessarily their first or strongest. And as Morgan argues

> ... in modern societies all of us are multilingual to some extent, participating in multiple discourses associated with the different practices we encounter in different parts of our life: within the family and with friends, at school and at work, engaging with the mass media in a variety of formal and informal social settings ... we also have different levels of fluency within these various discourses. (Morgan, 2007: 240)

Morgan uses the term discourse here to describe the patterned uses of language and other forms of communication whose deployment identifies the user as belonging to a particular community at a particular time and in a particular setting. The notion of discourse reminds us that language is not just a means of communicating our thoughts to other individuals but also expresses and shapes our identities, which is sensitive to the specific context in which it is produced, and may be used to control access to particular communities.

While researchers and educators have described some of the 'peculiarities' of mathematical language (e.g. Barwell *et al.*, 2007; Barwell *et al.*, 2002; Monaghan, 2000; Pimm, 1987), it remains the case that efforts to apply their insights in the classroom have largely been at a local level with individual teachers or teams of teachers seeking to address the issue in their particular setting, rather than there having been a more systematic approach to curriculum design informed by linguistic analysis. In this chapter, I will describe and illustrate one linguistic approach, corpus analysis, that, together with a software tool called a concordancer, can be used to map the mathematical langscape and so contribute to curriculum design. First, however, I will illustrate the multilingual nature of mathematics classrooms in London, United Kingdom, in order to give some context to these ideas.

The Mathematical Langscape in a Multilingual School in London

In London, 53.4% of primary school students and 49.3% of secondary school students are bilingual (DCSF, 2007), meaning that they use at least one language in addition to English. Typically, students' main languages come from a wide range, often a dozen or more in London schools, while teachers are most commonly monolingual in English, with perhaps some knowledge of a European language. The data drawn on in this chapter come from an ongoing research project[1] 'Dialogic teaching and learning with EAL pupils', a continuation of previous research into exploratory talk in mathematics (Mercer & Sams, 2006; Monaghan, 2005). The project focuses on how dialogic teaching can be used to support and enhance the performance of learners of English as an additional language (EAL) in mathematics. The 12 students taking part in the research project are from Years 10 and 11 (15–16 years old) and are learners of EAL.

The school participating in the project is a mixed-gender, multilingual, multiethnic comprehensive school in inner London serving students between the ages of 11 and 18. The approximately 1200 students come from a wide variety of ethnic backgrounds, as shown in Table 2.1.

Some 80% of the students in the school are from minority ethnic groups. Approximately 46% of these students are learners of EAL. The major languages spoken at home, other than English, are Gujarati, Urdu and Arabic. The school is representative of its surrounding mixed inner city community, with a diversity of class, ethnicity, faith and attainment. There is a significant gender imbalance, currently 42% girls to 58% boys. The proportion of students who receive free school meals is 29%, an indication of social deprivation that puts the school in the highest percentile nationally.

The students involved in the project are, in some senses, emblematic of the political and economic history that has led to the particular multilingual make-up of the United Kingdom. One student is a second-generation Pakistani, typical of many of the Asian students in the school whose families have a right to settle in this country arising from Britain's colonial past. He describes his parents as uneducated and English is only spoken at home between him and his siblings. He himself speaks English fluently and with a local accent. Another student is from Nepal and part of a relatively new group of immigrants to the United Kingdom with entitlement

Table 2.1 Ethnic backgrounds of students of a London secondary school

Ethnic category	*Of which ...*	*Of which ...*	*Of which ...*
White – 30%	British – 21%	Irish – 1%	Other – 8%
Black – 24%	Caribbean – 13%	African – 8%	Other – 2%
Mixed – 14%	White/Caribbean – 6%	White/African – 2%	Other – 6%
Asian – 23%	Indian – 12%	Pakistani – 8%	Other – 3%
Chinese – 0.3%			
Other – 8%			
No response – 0.7%			

Mapping the Mathematical Langscape

to settle here due to the part played by their country's soldiers in the Second World War recently being recognised by the British government. This student attended English-medium schools in Nepal from the age of three, but, while he has a good superficial control of the language, his accent and grammatical accuracy mark him as a recent arrival to the United Kingdom. A third student is from Iran and represents a significant number of students in London schools from the Middle East, who are mostly refugees. She herself is not from a refugee family, but is from one of a smaller number of families whose parents are working temporarily in the United Kingdom on diplomatic or commercial postings. Well educated in her own country, she has learnt English as a foreign language in school before her recent arrival here and is still reticent about her competence in the language, though very confident about her abilities in mathematics. A fourth student is from Slovakia, one of the countries in the enlarged European Union whose citizens now enjoy the right to work and live in the United Kingdom. His family has been in London for the past four years following his father's posting here as a diplomat. He previously lived in Denmark for three years, where he attended an English-medium international school and he has also learnt English as a foreign language. He has a very strong command of the language and his accent is a mix of Slovakian and American. London classrooms commonly reflect the dynamic mix implied by these brief portraits.

As part of the project I interviewed V, the 15-year-old Nepalese student described above. The school strives to provide an inclusive environment that respects and promotes diversity. This is reflected in comments made by V, who had arrived eight months earlier, when asked about the use of his home language in school:

FM: Have you been able to use Nepali in, in the school at all?
V: Yeah I got my friend P. We speak Nepali. It's easy to communicate.
FM: Ok. So, did you meet him here then?
V: Yeah.
FM: ... but is it used at all in lessons as apart from talking with P? ... Is that just you and P then?
V: Yeah.
FM: Ok. So you talk about the work mainly.
V: Yeah, we talk about the work.
FM: Ok. And what ..., how do the teachers react to that?
V: Ah, nicely.
FM: Good. Ok. So, there isn't, there isn't a problem about you speaking Nepali to each other? Nobody says no, no, no, you should be speaking English.
V: It's like that, they encourage us.

The teachers appear to have positive attitudes towards the use of the students' home language and seek to encourage it. The students themselves, however, do not necessarily regard this as being particularly salient. V's use of Nepali now that he is in England is not limited to exchanges with his friend; however, he also receives support from his older brother:

FM: So, your brother has helped you. So how ..., and that's in English as well then or not?
V: In English.
FM: Ok. Always in English? He doesn't just do it in Nepali?
V: No, no, in Nepali. Sometime in English, sometime Nepali.
FM: Ok. Does that help, being able to use both languages?
V: Yeah, yeah.
FM: Why?
V: Yeah, yeah, it is practice for me to speak English and more fun.
FM: But what's the Nepali then, why is that helping?
V: To speak English?
FM: No, to speak in Nepali.
V: Nepali, everybody understands Nepali, so it's easy to teach other.

This valuing of using both Nepali and English is at odds with V's general perception of the role of English in his education and, in particular, of his assessment of the kind of help from which he would benefit. V regards the problem as being entirely at the word level. I asked him what difficulties he had in doing more extended writing in mathematics, such as, for example, in coursework:

FM: ... what is difficult about it?
V: English. Because there is too many difficult sen- ..., difficult word, like ... some things are so difficult so we don't understand.
FM: But what kind of difficulty? Words or spelling or sentence writing?
V: Words, words.
FM: Just single words?
V: Yeah.
FM: OK.
V: We don't understand what's that mean so we just quit.
FM: So how do you find out the meaning of words?
V: From dictionaries.

Whenever I probed further on sources of help he would mention only dictionaries. This proved to be a very resilient attitude even in the face of other evidence that emerged about his experience of using English in mathematics and in other subjects. V's teacher had reported that V was particularly fond of algebra but was resistant to doing activities that

Mapping the Mathematical Langscape

involved more language. I asked V about algebra, in order to see if he was mindful of this preference:

FM: ... when you're doing algebra, there isn't much language in the questions.
V: Yeah, in algebra, so it's easy.
FM: When you're doing ratios, the questions are about words, aren't they?
V: Yeah, it's all the words, yeah.
FM: Well, could that be part of the problem for you?
V: Yeah. For language, yeah.
FM: Yeah. Ok. So what could be done about that?
V: A dictionary.
FM: You think that would solve it, just having a dictionary?
V: Yeah.

In spite of his recognition of the support he received from his Nepali classmate P and his brother, he did not think that more support from bilingual teachers or classroom assistants in lessons would be helpful:

FM: Would it have helped to have support or not?
V: Dictionary.
FM: Well, just a dictionary, that's ..., all you want is a dictionary.
V: Yeah, only dictionary.

It seems that for V, the major difficulty he faces as a learner of English in the mathematics classroom is 'words', and that the main support he needs to deal with this difficulty is a dictionary. While vocabulary may be the salient issue for V, my observations working as an EAL teacher in the school suggest that several other aspects of language can be problematic. Often these difficulties arise from ambiguities between 'mathematical English' (the kind of 'peculiar' English Halliday, *op. cit.*, was talking about) and 'ordinary English' (as spoken in everyday life). In this chapter, in addition to vocabulary, I will focus on syntax.

Vocabulary

Perhaps the most obvious level at which the ambiguity of mathematical English becomes evident is that of vocabulary. For example a word like 'similar' has a more precise meaning when applied to objects in the mathematics classroom (corresponding angles are invariant and all side lengths vary by the same scale factor) than in the world beyond (see Jones, this volume, for an example). On one occasion, when working with a new arrival to the United Kingdom on an activity that involved sorting six cut-out triangles into two sets of 'similar' triangles, I realised that he was having problems in understanding what he needed to do. I asked him what 'similar' meant. His response was 'the same' and, as they all looked pretty much the same, he was understandably confused. He had also understood the task to mean that he had to put the triangles into pairs – sets of two rather than two sets. These marked uses of mathematical English continue at other levels.

Syntax

In mathematical English, we say 'three quarters *is* bigger than five eighths' rather than 'are bigger' because fractions are treated as single entities, so, in spite of quarters being plural, we use the singular form of the verb. Teachers of EAL have to explain to students whom they have diligently reminded of the need for subject–verb agreement in English why, in mathematics, it is sometimes different. Such syntactical confusion can also arise at other levels in mathematics, notably in algebra with the use of brackets. Another new arrival, working on a mapping problem, translated 'w $\to$ 2(w–6)' as 'multiply by 2 then subtract 6', following the order of physical presentation on the page rather than the conventions of operations. This type of error, of course, is not restricted to bilingual learners, but is likely to be more common due to the dual burden of processing not only mathematics but also its translation into a second language. The point is that a heightened awareness of language issues will not only be supportive of EAL students but of all students.

It was from experiences such as those described above that I began to wonder about the relationship between language and mathematics and how one might track and make explicit the generally hidden language syllabus that is inevitably contained within the mathematics curriculum. The development of mathematical corpora and use of concordancing software to interrogate them now make Halliday's (1975) goal more realistically attainable and it is to this that I now turn.

Mapping the Mathematical Langscape: Corpora and Concordancers

A *corpus* is a body of authentic language data. These data can be written or spoken text. Corpora now tend to be stored electronically and available for interrogation using software tools. Corpora may be large or small, depending on their purpose. The large British National Corpus, for example, consists of some 100 million words of written and spoken English compiled since the early 1990s.

The use of corpora in language research, as with any method, has its limitations, which should be acknowledged at the outset. For example a corpus cannot show anything more than its own contents and so must always remain a partial representation of the language. Any generalizations or conclusions drawn from the study of corpus evidence must be viewed as deductions rather than facts and depend on the quality of interpretation and intuition of the researcher (Hunston, 2002). A further difficulty is that all texts are inherently multimodal and created within and for a given context. There is always a visual aspect to a written text produced by such aspects as the choice of font, paper, binding and so on and these aspects are lost in a corpus, which deals with plain text stripped of its other features (Kress & van Leeuwen, 1994; O'Halloran, 1998). This point also applies to spoken texts, since recordings cannot convey any gestures or spatial organization of the speakers that also impact on the use and meaning of language in a given context. These limitations, however, are powerfully counter-balanced by the externally examinable evidence of the corpus, in contrast to the reliance on the researcher's intuition about the contents of their own and others' minds that beset other linguistic research. It is, however, important to recognise and acknowledge that the use of corpora in language study is only one tool and needs to be used in conjunction with other observations.

A *concordancer* is a piece of software that enables its user to track and display patterns in a body of language by displaying key words in context (KWiC), as illustrated in Figure 2.1. The word under investigation is displayed in bold in the centre with a defined number of characters

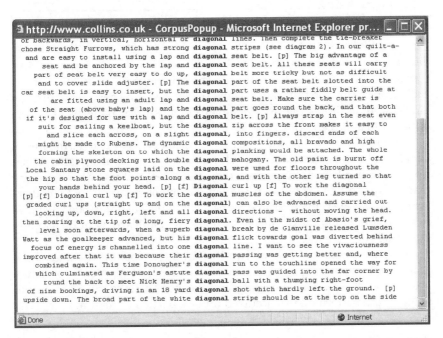

Figure 2.1 Screenshot of Cobuild corpus illustrating use of 'diagonal' in general use

Table 2.2 Frequency table of words from the field of 'area, shape and volume'

	Area, shape and volume by frequency	*Frequency*	*Area, shape and volume, alphabetical*	*Frequency*
1	Area	190	About	32
2	Draw	126	Above	6
3	Tiles	124	Accurate	3
4	Squares	114	Accurately	7
5	Shape	106	Acute	2
6	Triangle	105	Added	3
7	Rectangle	76	Adjacent	3
8	Fit	65	Angle	22
9	Pieces	65	Angled	16
10	Height	61	Angles	53
11	Figure	57	Arc	1
12	Base	56	Area	190
13	Different	56	Areas	32
14	Sides	56	Around	3

displayed on either side of it to allow the context to be clarified. Most concordancers will take the reader to the original text when the word is clicked on so that the full context can be examined. This method of display is useful as it allows the investigator to identify real patterns of use as opposed to assumed ones.

Another way of displaying text in a concordancer is by using a frequency table, as illustrated in Table 2.2, which displays vocabulary items taken from the topic of 'area, shape and volume' in a set of mathematics learning materials. The table shows, on the left, the most common mathematical terms and, on the right, the same terms in alphabetical order. Such displays can be used to identify key terms for teaching and learning or developing glossaries for students.

This sort of display also allows the reader to look at distinguishing patterns of use or simply to compare frequencies of different words in a text. It would require further analysis of the word in its contexts to uncover whether the patterning has occurred for any particular reason.

Using a concordancer with mathematical texts

Concordancers have been used to analyse all manners of texts, literature, journalism, legal proceedings, school history, written and spoken.

As a result, we have gained insights into how such texts are structured and used to create meanings. A corpus of school mathematics texts might similarly give us valuable insights into the language of mathematics and how students might be enabled to become more effective users of both mathematics and language.

The mathematics department in V's school operates an individualised learning scheme from Years 7 to 10 (11–15 years old) in which students are taught in mixed-attainment classes and set individual programmes of work within a particular module (such as 'shape and space' or 'data handling') according to their current level of ability. This approach is supported through the use of a mathematics scheme called Secondary Mathematics Individualized Learning Experience (SMILE), which consists of work cards, work sheets, computer activities, puzzles and games, and is premised on a problem-solving approach in which students are expected to take responsibility for their own learning.

I created a corpus of mathematics materials based on the SMILE resources in use, some 1318 activities of the 1418 that were available at the time of compilation. This corpus accounted for approximately 93% of the SMILE scheme, the remaining 7% being either unavailable or held in an unsuitable format (i.e. computer activities without any written text or activities that involved no written text). It seems reasonable to argue that, as the corpus is based on such a high percentage of the materials in the SMILE scheme and covers all levels of the English National Curriculum in mathematics, it serves as a sound basis for the analysis of mathematical English as developed through this particular scheme. In the following section, I will give a couple of examples of the sorts of things that can be done using a concordancer.

Exploring Vocabulary: What is a Diagonal?

According to one mathematical dictionary:

> A diagonal is a straight line drawn from one vertex of a polygon (or solid figure) to another vertex. (Abdelnoor, 1979: 28)

According to the Collins Cobuild dictionary:

> A diagonal line goes in a slanting direction away from another line.

The sample below is taken from a large corpus of English (the Cobuild corpus), which contains some 56 million words based on written and spoken texts from a wide variety of genres, including newspapers, novels, conversations, etc.). The sample shown in Figure 2.1 is only a small extract but it is interesting to note how a diagonal is used here very frequently in the context of sports – mostly football (soccer) and rugby.

In a study of written texts in the secondary mathematics classroom (Monaghan, 1997, 1999), I explored the occurrence of the word 'diagonal'

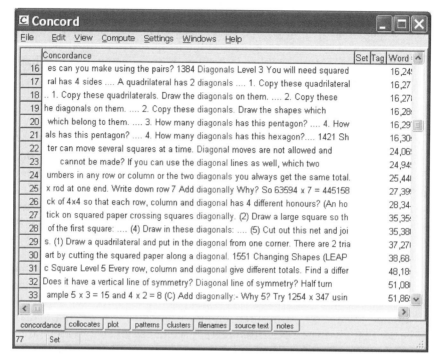

Figure 2.2 Use of 'diagonal' in its various forms in SMILE materials

in the SMILE mathematics scheme and the ways in which the two distinct meanings apparent in the above quotes were reflected in it (Figure 2.2).

Analysis of the word (or in linguistic parlance 'lemma' that includes its multiple forms 'diagonal', 'diagonals' and 'diagonally') revealed that 'diagonal' appeared in the SMILE scheme as:

- an adjective, 'diagonal moves are not allowed';
- a noun (singular and plural), 'the vector OR is the diagonal of a box ...';
- and as an adverb, 'you may not hop diagonally'.

This analysis showed that the word was used with both its mathematical meaning (as an attribute) and its ordinary meaning (indicating orientation). By looking at the relevant frequency of occurrence, I was able to discover some other interesting features. The corpus was structured in such a way that I was able to see at which National Curriculum levels the word appeared and so track its development through the SMILE materials (the English National Curriculum is organised into eight levels).

The word 'diagonal' did not appear at National Curriculum Level 2 in the scheme, suggesting that there might be a need or opportunity to plan another activity to fill this gap. Further, there were particular clusters where it seemed to feature more heavily than at others (i.e. between Levels 4 and 7), which may reflect its place within the hierarchy of mathematical concepts. What was most interesting, however, was that the lemma occurred 66 times across 35 separate activities throughout the scheme and of these 32 were in its technical, mathematical sense and 34 were in its non-mathematical sense. This ran counter to my intuition that a mathematics scheme would privilege mathematical meanings over ordinary meanings. That it does not has potentially serious implications for how students like V experience and understand mathematical terminology, since the most common form they encounter is likely to have the most salience.

Exploring Multimodal Texts

There are occasions where the multi-modal nature of mathematical texts can support bilingual students by allowing them to draw on visual aspects when their limited English leaves them unable to understand the written text. Let us take, for example, the case of J, the 15-year-old Iranian student, mentioned earlier. J had arrived in the United Kingdom not speaking any English only six months before I began working with her class. Despite this, her teacher recognised from J's computational abilities that she was an able mathematician.

J was working on an activity involving rotation (shown in Figure 2.3). The written instructions asked the student to 'draw axes with x values from −6 to 6 and y values from 0 to 6' and 'plot points at A (1, 2), B (6, 1) and C (5, 6)'. J had successfully done this, but, when I asked her to read the instruction, she told me that she did not know what the words 'plot', 'point' or 'axes' meant. She had been able to do what was asked because the illustration on the card had made it obvious to her what needed to be done.

What was interesting, however, was that she had started not from the original triangle (ABC) but with the rotation ($A_1B_1C_1$). In doing so, she was probably following the normal left–right writing direction of English, even though she herself is a Farsi-speaker in which text is written from right to left. On the one hand, this can be taken as evidence of her developing English literacy and on the other, evidence of an area for development in her mathematical English literacy, or her multi-modal literacy, in which conventions of one system may not apply within another system. As Street points out, in mathematics

> the issue of modality is central since numerical principles and procedures are always represented in a variety of modes – from oral and written (using symbols) to visual (in mathematics education terms 'iconic'), including layout and ordering as in a number square or number line, to actional (or 'enactive'), as in the use of concrete

ROTATION

Smile 1112

1. a) Draw axes with *x* values from -6 to 6 and *y* values from 0 to 6.
    Plot points at **A** (1, 2), **B** (6, 1) and **C** (5, 6). Join them to form the triangle **ABC**.
    Trace the axes and the triangle.
    Rotate the tracing paper through 90° *anti-clockwise*, about (0, 0).
    Draw the rotated triangle, label it $A_1B_1C_1$.

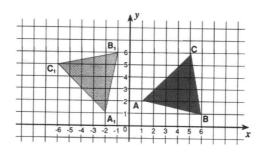

b) Copy and complete the mapping:

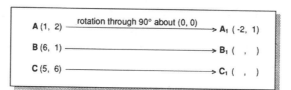

Figure 2.3 Smile Card 1112 – rotation
Source: © Royal Borough of Kensington and Chelsea

apparatus for number or as in the movements across diagrams. (Street, 2006: 219)

J's 'error' allowed me to have a conversation with her about the use of subscript numbers to indicate a rotation, an important part of her mathematical literacy. In part (b) of the exercise, J did not initially copy the model of writing the description of the mapping along the line as exemplified in the illustration above. For her, it was sufficient to write the mapping alone. I was able to persuade her that it was useful for her to practise the use of the description in order to focus on the particular uses of the prepositions 'through' and 'about', with which she had not been familiar. Research has shown that, compared with native speakers, second-language learners

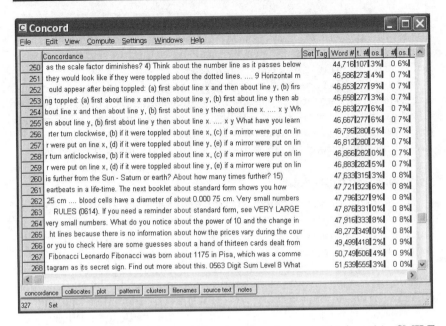

Figure 2.4 Screenshot of concordancer showing use of 'about' in SMILE mathematics materials

typically use fewer of the lengthy noun phrases (frequently involving prepositions, as in this example) essential to effective academic writing (Cameron & Besser, 2004; Halliday & Martin, 1993). The corpus can be helpful in identifying opportunities for learners to develop their language use in this area in a meaningful mathematical context. In the case of 'about', for example, I was able to create a sub-corpus based just on those activities from the field of motion geometry and then identify uses of 'about' within that area. The screenshot shown in Figure 2.4 is a sample of usages of 'about', illustrating that they are not only confined to the mathematical meaning relating to the centre of rotation but also appear in more 'normal' uses. As with 'diagonal', the everyday meaning of a word, even in resources aimed at teaching a specialist register such as mathematics, is likely to be more prevalent (and therefore reinforced) than the specialist mathematical meaning of the word. (See Halai, this volume, for a cross-linguistic example of the same issue.) Note also the varied syntax that can be seen in these usages.

Developing Resources

Analysis is one thing, classroom practice another. There is insufficient space in this chapter to do any more than provide an example and some discussion of materials designed to foster learning mathematics through language and vice versa. That said, it is important to emphasise that language and content are not separate entities but rather different lenses through which learners and teachers can be helped to see more clearly how learning and teaching can best progress.

In essence, there are three approaches we can take: we can use existing materials as they are and exploit them for their language potential, we can adapt existing materials in order to bring out latent language development opportunities or we can create new activities to meet our learners' needs. I will limit myself to giving examples from the two ends of the continuum.

Using existing materials

The genesis of the example shown in Figure 2.5, 'Bengali Number Jigsaw', is interesting. The jigsaw was initially devised by an 11-year-old student who had recently arrived from Bangladesh and who had very little proficiency in English. He did a similar jigsaw using Hindu-Arabic numerals and was asked by his mathematics teacher to produce a Bengali version. This version was then produced commercially by SMILE. The effect in the classroom was very powerful since, mathematically speaking, it is a relatively simple task for someone who knows Bengali. When given to students familiar only with Hindu-Arabic numerals, it became much more challenging and resulted in other Bangladeshi students being approached for help as 'experts', thus raising their status in the classroom, and their self-esteem, a known affective factor in language acquisition.

Figure 2.5 SMILE 1858: Number jigsaw originally developed by a 'new arrival' bilingual student
Source: © Royal Borough of Kensington and Chelsea

Mapping the Mathematical Langscape 29

Creating new resources

In the activity shown in Figure 2.6, students are provided with and expected to make sense and use of a specifically mathematical text type (the pie chart) and the language associated with it. Analysis of the resources available had demonstrated that the language of pie charts and other forms of data representation was not explicitly developed. This activity was designed to help 'fill the gap'. Only a close reading of all aspects of

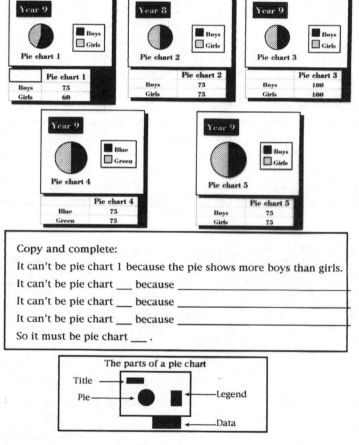

Figure 2.6 Pie-chart activity

such a diagram (the title, legend, pie and data) will provide students with the correct solution. The writing task is structured as a straightforward gap-fill with a simple sentence structure modelled in the example. It also makes explicit the elements of a specifically mathematical form of visual presentation, deploys a common mathematical method of proof by elimination and gives practice in using a logical connective ('because'), which bilingual students have been shown to have difficulties in using (Dawe, 1983). It is impossible to get the right answer and complete the task without understanding and deploying particular features of mathematical English. The activity represents a sound marriage of language and mathematics, and scaffolds the student's engagement with both.

Implications and Future Directions

A crucial challenge for teachers and researchers working in multilingual settings is to identify gaps and opportunities for fresh interventions that help our students learn what they need to learn in a more linguistically principled way. A further ambition, with the needs of students learning mathematics through EAL in mind, is to develop a language curriculum through mathematics. Sinclair and Renouf (1988) and Willis (1990) have argued for the development of a 'lexical syllabus', which does not just mean a focus on lexis (vocabulary) but on all aspects of language, with lexis as the organising principle. Sinclair and Renouf suggest that the syllabus should focus on (a) the most common word forms in the language, (b) the central patterns of usage and (c) the combinations which they usually form. Clearly, a well-developed corpus and the use of concordancing software provide a potential basis for drawing up such a syllabus. Willis, in fact, argues that a corpus can be the syllabus since, if it contains samples

Table 2.3 Reflections of broader culture in SMILE materials

Place	*Name*	*Maths*	*Jobs*	*Food*
Africa	Adam	Abacus	Artist	Bacon
America	Amandeep	Archimedes	Builder	Banana
Anytown	Eve	Chisolm	Mathematician	Coke
Babylonian	Jothi	Diophantine	Policeman	Milkshake
Bangladesh	Kitty	Euclid	Porter	Orangeade
Brainage	Latifa	Euler	Schoolkeeper	Pizza
Britain	Laxmi	Fibonacci	Shopkeeper	Quiche
Mayan	Mee Fing	Gelosia	Teacher	Sausage

of the most frequent patterns of the most common words, that will be sufficient to exemplify what learners need to learn. The task of the teacher and/or researcher is to find ways of enabling learners to engage with the corpus in meaningful ways.

To address broader aspects of the mathematical langscape, the analysis would need to be extended to include, for example, affective aspects of the curriculum, such as how it reflects cultural diversity – what kinds of names appear, what jobs people have compared to the mathematics they use, how disability is reflected in the materials and so on. Table 2.3 contains some examples (from Monaghan, 1997) that reflect aspects of diversity that appear in the SMILE resources, although what is also needed is an examination of these in context, to see whether they reflect positive or negative images, for example. With appropriate tagging of the corpus such analyses are possible.

It is through close examination of the particular in the mathematical langscape that we may illuminate the peculiar and render it meaningful. The development of a representative corpus of mathematical language, in my view, offers the best chance of developing a linguistically principled mathematics curriculum.

Note

1. The research is funded by the National Centre for Excellence in Teaching Mathematics (reference G071206) and runs from March 2008 until July 2009.

Chapter 3
Somali Mathematics Terminology: A Community Exploration of Mathematics and Culture

SUSAN STAATS

Teachers in many parts of the world work in multilingual classrooms, increasingly so due to migrations caused by violence and by shifts in political and economic structures (Gutstein, 2007). Multilingual classrooms are complex, bringing together students with educational needs that, taken individually, would call for very different interventions: teachers and students who share, or who do not share, a common language; students whose families hope to return home and those that do not and very young children or teenage students who do not speak the language of instruction. My own algebra classes at the University of Minnesota serve many Latino, Hmong and Somali immigrant students who are recent graduates from local secondary schools. The estimated five million language learners in United States primary and secondary classrooms (Moschkovich, 2002: 190) represent dozens of languages, few of which are shared by their teachers.

This variety of multilingual situations raises significant pedagogical challenges for teachers (Morgan, 2007). A teacher who does not share a home language with her students cannot, for example, use the informality of the home language to develop rapport with students. While the teacher may encourage students to discuss task scenarios in their home language, she cannot sense when the students touch upon a mathematical insight or when they need to develop greater contextual knowledge. The teacher cannot codeswitch to emphasize transitions between different pedagogical purposes such as explanation and modelling formal mathematical speech. This is a common scenario globally, and it would seem, one that limits a teacher's options for an intentional pedagogy that engages students' home languages as learning resources. There is a need, therefore, for case studies of pedagogical action and curriculum design that teachers can implement in multilingual but linguistically distinct classrooms.

In this chapter, I describe a community-based project centred on mathematics vocabulary in the Somali language that addresses these issues. The project was inspired by a conversation with one of my first-year undergraduate algebra students in the spring of 2003. He submitted a homework paper headed *'dheeli toosan'*, with the subheading 'linear inequality'. He learned the phrase while being tutored by his cousin, who had taught secondary school mathematics and science in Somalia around 1990 and who is well acquainted with Somali mathematical terminology. I asked my student if he had ever heard the phrase before he learned it from his cousin. He thought doubtfully for a while but then realized that he knew the word *dheeli* as a term for an unbalanced placement of a load on a camel or donkey. Later, I asked a female student if she had ever heard of a word for 'inequality' in Somali. Not surprisingly, she had not, but when asked if she knew of any word with a similar sound, she recounted conversations with her mother in which she was about to leave the house with her *xijaab* hair covering poorly tied. She suggested that *dheeli* could also refer to a situation in which one's life is out of balance. In this way, I began to think of *dheeli* as a term that could be glossed as unbalanced, but that could have somewhat different connotations based on gender.

As a cultural anthropologist, I have recorded and translated religious texts in indigenous communities in Guyana, South America. I was curious to find out if I could use the methods that an anthropologist uses to learn and describe an unwritten language as a means of engaging bilingual students in the mathematics classroom. In this chapter, I give an overview of educational issues that Somali immigrants face, outline the process of preparing for and conducting linguistic interviews on mathematics terminology, and describe ways in which Somali students create mathematical understandings by considering the cultural meanings of Somali mathematical vocabulary. My intention is to stimulate dialogue on two issues. First, I wish to contribute a case study for debate on the relationships of culture and mathematics; in particular, on the question of how cultural we wish our mathematics to be. Second, I outline aspects of Somali educational history in order to evaluate the possibility of conducting community-based studies of mathematical terminology for other languages. In some respects, the Somali project described here may serve as one model for teachers who wish to involve their students' languages in the classroom. In other respects, the unique history of the Somali language and of the immigrant community in Minnesota structured the project in ways that may not transfer to other contexts. Nevertheless, the discussion offers a starting point for teachers who wish to establish their own community language projects to enhance mathematics learning.

Mathematics Terminology as Metaphor

Prior to the 1970s, primary schooling in Somalia was conducted in Arabic and Italian, and secondary schooling was conducted in English. In 1972, President Maxamed Siyad Barre selected the English alphabet as the first script for the Somali language. Somali became the mandated language of instruction and over the next decade, a series of Language Commission committees constructed a vocabulary for mathematics and science using terms associated with the nomadic and coastal economies (Abdi, 1998; Jama Musse, 1998; Johnson, 2006). The vocabulary was used in textbooks and classroom instructions from elementary grades through to university-level classes.

The resonant names of Somali cultural objects and actions were selected to represent the abstract, limited and technical concepts of mathematics and science. For example the origin of a coordinate system is termed *unug*, 'where something is from, the first blood, the seed, the beginning', as a Somali undergraduate student phrased it. An ordered pair is *lammaane horsan*, in which *lammaane* means twin, or people from the same group who share a way of thinking or feeling. Because the Somali mathematics vocabulary was constructed from words used in daily life, each word is a *conceptual metaphor* (Lakoff & Núñez, 2000: 42). A conceptual metaphor is a mapping from a concrete conceptual domain (such as agriculture) to an abstract one (in this case, mathematics) that preserves some elements of inference and that can facilitate mathematical thinking. The concrete circumstances of a beginning point and of two elements united into a single meaningful element express the abstract mathematical relationships to which they correspond in relatively simple and direct terms. Metaphorical connections between familiar objects and mathematical ideas have been a basis for developing mathematics vocabulary elsewhere, for example, among Māori (Barton *et al.*, 1998).

As metaphors, many mathematics terms in Somali are based on visual similarities to daily objects. The mathematical parentheses symbols '( )' are represented by the word *bil*, crescent moon. The symbol '(' is read *bil-furan*, moon-open, and the symbol ')' is read *bil-xiran*, moon-closed. Alternatively, parentheses can be read as *qaanso*, the bow that shoots an arrow. In some cases, the visual metaphor connecting daily life and mathematical vocabulary acts in a similar way to English mathematical terms. The word intersection, in both English and in the Somali term *isgoys*, can mean the place where two roads meet or, in mathematics, the crossing of two graphs. Establishing an association between a culturally important object and a mathematical concept based on similarity in shape is known as iconicity, so that the daily object acts as an icon or an image for the mathematical idea.

The connection between daily life and mathematical terms can occur through a slightly different type of iconicity, too, taking the name of an action or a gesture to represent a mathematical object. A line in Somali is *xariiq*; a horizontal line is *xariiq jifta*, line asleep, and a vertical line is *xariiq taggan*, line standing up. *Jibbaar*, exponent, takes its name from a measuring technique in which a length of string is held to a desired length, which is transferred head-to-tail across an object so that integer multiples of the length can be obtained. The action of laying the length in sequence is reminiscent of writing a base a given number of times.

Viewed historically, the metaphorical basis of Somali mathematics words is not especially different from that of English ones. Many English mathematics words are based on terms from daily life: consider the *face* of a polyhedron. In English, however, the history of mathematical vocabulary is so long that, in many cases, the original metaphor is no longer apparent to contemporary users (Núñez, 2006: 173). In Latin, for example, *matrix* meant a pregnant woman, so that the capacity to generate something new was adopted as a mathematical expression (Schwartzman, 1994: 132). *Calculate* comes from a Latin term for chalk or limestone, referring to the use of pebble and stone boards for arithmetic (Schwartzman, 1994: 39). *Polygon* uses the Greek root for knee to represent the angles of the figures; in a similar manner, Somali uses *geeso*, horn, for vertex.

In a few cases, the experiential meaning is still associated with an English mathematical word. Sfard compares the symbolic potential of the English terms slope and derivative, and suggests that accessible, ordinary meanings can support mathematical understanding:

> More often than not, the signifiers themselves become more or less powerful meaning activators, with their signifying strength depending on their history. Thus the name *slope* obviously brings a much richer semantic heritage than the word *derivative*. (Sfard, 2000: 76, original emphasis)

Although many English terms have their origins in a conceptual metaphor, the concrete domain that formed the basis of the metaphor has been lost, and the English term is often an arbitrary string of sounds to mathematics students. Somali immigrant students, however, have the opportunity to regain the semantic heritage of their own mathematical words, if teachers and community members work together to develop this knowledge base.

Educational Issues for Somali Immigrants

Community-based research requires teachers to develop a greater understanding of students and their communities than they may encounter in everyday professional life. A teacher who studies mathematical aspects of students' language must learn some of the history of the community, the circumstances that bring teachers, students and parents together. I report in this section on specific issues facing Somali students in Minnesota. Teachers who wish to inventory mathematical or linguistic resources in their own students' communities will need to develop similar types of knowledge for their own case studies. Besides providing a basis for competent and respectful conversation, this background will help teachers identify educational resources in the community. Immigrant parents may have mathematically relevant knowledge based on culture or work experiences that can become a pathway for contributions to their students' educational experiences and involvement in unfamiliar institutional educational settings. Immigrant families may be undergoing shifts in daily-life activities and in authority that create divisions between adults and children, so that these conduits of educational cooperation are even more important. The Somali case described below is unique, but the principle of becoming aware of history, cultural challenges and potential mathematical learning resources within immigrant families is a necessary preparation for teachers doing community-engaged work.

Somalis began migrating to Minnesota after the collapse of the Somali government in 1991. Many families sought refuge in neighbouring countries, especially in Kenya. Formal education was not always available there to young people with refugee status and, as a consequence, many of the young Somalis who immigrate to the United States have had few, if any, years of formal schooling. The twin cities of Minneapolis and St Paul, Minnesota, are now home to the largest Somali community in the United States. As a result of their interrupted education, few Somali students are familiar with the mathematics and science terminology of their language. They have become separated from a potential learning resource as they move through the American education system.

At the beginning of the civil war in Somalia, educational facilities were among the earliest targets destroyed; educated Somalis were frequent targets of factions in conflict with each other (Abdi, 1998: 335–336). Adult Somali refugees in Minnesota include many highly educated people whose credentials were lost, intentionally destroyed for personal safety or that go unrecognized in their host country. Some of these men and women apply their learning through the educational system as teachers, translators and educational assistants. A former professor and industrial scientist from Somali National University, for example, now teaches mathematics and science in a local school. Somalis who work in the Minnesota education system recall much of the technical vocabulary from their own schooling in Somalia. Even so, most Somali students in the United States learn from English-speaking teachers.

Adapting to life in the United States presents serious personal, social and educational challenges for Somali youth. Halcón *et al.* (2004) report that over 9% of Somali youth aged 18–25 years in Minnesota have endured war trauma and torture resulting in depression, anxiety and sleep disorders, although many cope in positive ways through prayer or social support activities. About 80% of Somali youth in Minnesota entered the United States unaccompanied by their families; many entered in their upper teenage years. Even when members of an extended family are re-united in Minnesota, laws limiting apartment occupancy result in elders living separately from the rest of their families, so that elders' traditional authority roles as supportive guides and family dispute mediators are weakened.

In another study (Robillos, 2001), Somalis in Minnesota identified education, along with employment and housing for their extended families, as their biggest concern. Study participants believe that Somali children's educational challenges are rooted in language. They noted that grade-level placement of immigrant students by age rather than by knowledge or by academic or language proficiency makes schooling ineffective. Parents are frustrated that language barriers prevent them from helping children with schoolwork. Adult Somalis report that children become exhausted struggling to master academic content, as well as spoken and written language, simultaneously. They suggest that access to computers, audio and video technology could assist students' learning (Robillos, 2001: 13–14). The study cites the 'enthusiastic willingness' of both children and adults to participate in English-language learning programmes (Robillos, 2001: 7).

Schools have not always been comfortable places for Somali students in the twin cities, particularly for girls (Robillos, 2001). The lack of gender differentiation in dress and in shared space among American boys and girls is disturbing to some Somali youth. Girls endure teasing by non-Somali boys, and, at the same time, girls who continue to wear the *xijaab* withstand criticism from Somali girls who have decided to forego it. The students surveyed in Robillos' study did not feel that these pressures affected school performance, but clearly, schools are not always emotionally safe places. It is notable that among the 18–25-year-old group, Somali women were far more likely than men to regret their decision to enter the United States (Halcón *et al.*, 2004). Only about 39% of the Somali youth in Halcón's survey had graduated from high school.

In many ways, Somalis' experiences in Minnesota mirror experiences in immigrant communities elsewhere in the world. In Helskinki, for example, as in the twin cities, many students immigrate as teenagers, the 'generation in-between' (Alitolppa-Niitamo, 2002: 275). These students must master academic content in a new language, begin to construct an adult identity in a new society and prepare for life after secondary school within the space of a few years. Parents in both cities value education highly, but they are frustrated with language barriers that prevent their involvement with their children's homework, with culturally based educational practices like co-educational physical education classes, and with the lack of Islamically oriented educational experiences.

Somali immigrant communities have been pro-active in creating self-help programmes like community schools and literacy programmes in Yemen (Waters & LeBlanc, 2005: 142), Helsinki (Alitolppa-Niitamo, 2002), Liverpool (Arthur, 2003) and in the twin cities. Schooling and after-school programmes in the Somali language help students maintain language skills and master academic content in a culturally sensitive setting. Somali language classes in Helsinki, for example, were more orderly, enjoyable and 'soothing' for Somali students than the transitional, Finnish language 'immigrant classes' (Alitolppa-Niitamo, 2002: 283).

These case studies suggest that Somali immigrants tend to value education and are willing to involve themselves in educational programmes. There is a desire for culturally appropriate educational experiences and for learning opportunities in the Somali language. People express the need for more interactions between immigrant and pre-existing communities (Alitolppa-Niitamo, 2002), creative, perhaps, technology-driven learning support (Robillos, 2001) and cross-generational interaction (Arthur, 2003). A community-based project on mathematics terminology addresses many of these interests. By spanning academic and cultural knowledge, such a project enhances language learning in both home and school languages, and fosters communication between elders and youth.

Instructor-Community Dialogues

A teacher who wishes to develop community-based research on mathematics words or on mathematics knowledge will face dilemmas at several different levels. Community participants are likely willing to negotiate the value, intent and audience for the work. Once knowledgeable community members agree to participate, a teacher will need to make moment-by-moment decisions about how to direct a conversation about mathematical terminology. Furthermore, the outcomes of this research will be determined by educational values of both the teacher and the participants. All participants are likely to question the degree to which cultural beliefs are relevant to mathematical learning. In this section, I describe some organizational aspects of the project referred to in this chapter so that interested teachers can begin to envision how they would engage with the dilemmas of community-based language research.

Finding contacts and preparing for interviews

In my own case, finding people in the Somali community in the twin cities to participate in interviews on mathematics terminology was relatively easy. The Confederation of Somali Communities (CSC) maintains an office in a community centre close to the University of Minnesota; the president of the CSC helped me to contact Somali teachers in the public school system and in a community school that provides supplemental instruction in Somali with an emphasis on mathematics, science and writing. Staff with a University of Minnesota English as an Additional Language programme, Commanding English, are well connected within the public schools, and they gave me contacts as well. The tutoring centre of the University of Minnesota Multicultural Center for Academic Excellence was also a resource, as they hire peer tutors with a variety of cultural and linguistic fluencies. Taken together, these facilities helped me find contacts through community links, through educational ones and through Somali undergraduates who had successfully and recently negotiated the public school system.

To prepare for interviews, I created a word list based on the introductory and intermediate algebra textbook that I use in my freshman classes. Later, I found a published bilingual list of Somali mathematics words (Haan Associates, 1992) and used this as an additional source for word lists. The Haan publication lists only the Somali words and their English equivalents. Because my main interest was to learn about the cultural and historical meanings of these words, this source did not compromise the interview process.

While all the teachers, community members and students who have participated in the interviews as consultants have been cordial and willing to contribute, they differ in the value they see in the project and the potential audience. Some Somali educators consider mathematics to be a 'pure' subject, best kept separate from the rest of the world. They see value in compiling bilingual dictionaries of mathematics terms, but, for them, the connection to culture is irrelevant to the study of mathematics. These educators consider the audience of the project to be university professors. Others see some value in the cultural and historical dimensions of the project, but believe that these associated meanings can confuse mathematical understanding. Still other educators support the use of Somali mathematics terminology and have initiated ambitious, independent projects to collect accurate terms and to develop exercise books around them. Some see the possibility of including cultural and historical definitions along with the mathematical ones as an option that will be intriguing for immigrant students, but that might not be important for students remaining in Somalia. In general, even people whose educational values and priorities differ from my own have expressed a willingness to contribute their knowledge of Somali mathematics terminology.

Questioning techniques

Words rarely translate from language to language in one-to-one correspondence. Somali mathematics words are somewhat unusual entities to translate, then, because the mathematical component of their meaning corresponds with the technical meaning of mathematics words in English or Italian. But the metaphorical basis of Somali mathematics terms means that they are polysemous: a word like *bil* is best understood as a single word with multiple, related meanings instead of two separate words that sound the same, one meaning moon the other meaning parentheses.

Given published resources and highly educated consultants, it has usually been fairly easy to determine the Somali equivalent for an English mathematical word. Most of my interviews, then, are directed towards eliciting cultural and historical meanings for the words. The overall structure of questioning is 'How do you say *inequality* in Somali? Does the word *dheeli / inequality* mean anything in daily life?' Some open-ended variations on this questioning have been helpful. 'Why do you think that word was chosen for that math concept? Is that word ever used in daily life?' Another variation that has sometimes been successful when others have failed is 'Do you know any word that sounds like *dheeli*?'

Conducting interviews

A linguistic interview is not a particularly natural form of interaction for most people, and consultants must probe and experiment to find out what kinds of information the researcher is seeking. At times, consultants' responses may seem long and tangential to the researcher's purpose. The researcher's response at these points can have a strong effect on the types of information that the consultant will subsequently offer. In the following selection, I demonstrate a combination of these attitudes towards a consultant. AA is a highly educated physical scientist who now works as a teacher in Minneapolis. This selection is about 45 min into our first meeting together. We both wrote and drew pictures as we spoke, but at Line 6, I took an opportunity to steer the conversation back to the term *bil*, moon, so that I could investigate some guesses about its meanings:

1   **AA:**   Then we have these parentheses, they call *bil. Bil* means the moon.
2   **SS:**   Oh, that's beautiful!
3   **AA:**   Yes. (Writing the symbols for open and closed parentheses.) This they call *bil furan* and this they call *bil xiran*. So now it's open and now it's closed.
4   **SS:**   Tell me if I'm right. I think traditionally in Somalia people knew a lot of astronomy. They watched the stars.
5   **AA:**   Wow, yes! Yes, where I came from, for example, I'm from Southern Somalia, and we have almost . . . 7 islands in the Indian Ocean, so always we trade with fish in Kenya so we have to look for the stars, they say every month, the first seven days, the first week, we have to wait until the waves come down and I think the last week of the month because

there is some. And so we give name to all the stars. There is stars that you see until 10 o'clock and there is some that are coming up at 4 o'clock in the morning and you can know (unintelligible). And then there is some stars that you know are always in the north and others are always in the south so you know

6   **SS:**   (interrupting at a slight pause) You always know where you are, which direction
7   **AA:**   (interrupting) Yes. That's how they use *bil*.
8   **SS:**   Now that *xiran*, without math, does that mean anything?
9   **AA:**   *Xir* means closed and *fur* means open.
10   **SS:**   And so it sounds like maybe not, when people were watching the skies for directions, they didn't
11   **AA:**   No, no, this was just for math, for math, for math. Now sometimes they have the ... meridian ... stars, or something like that and they call *furnareng* ... or something like that.
12   **SS:**   Now with *bil*, sometimes the moon looks like this or like this (drawing waxing and waning crescent moons)
13   **AA:**   Yes, yes!
14   **SS:**   And sometimes the moon is like this (drawing a circle for a full moon) and sometimes it looks like this (drawing a gibbous moon)
15   **AA:**   No, the one like *bil furan* (pointing to the waning crescent moon, shaped like the open parentheses symbol) for the first five days, but this one they call circle (pointing to the picture of a full moon)
16   **SS:**   Oh, how do you say circle?
17   **AA:**   *Wareeq*.
18   **SS:**   Oh that's it, after the dark moon, the new moon, we say,
19   **AA:**   Yes, after the new moon it's like this one on the fifteenth, and this one they call *tayeh*.

The speaker's description at Line 5 was exactly the type of discourse that I was seeking. And clearly, this speaker has a great deal more to say on Somali scientific and mathematical knowledge of astronomy. But still, I knew that the interview period was coming to an end and I wanted to understand better why *bil* was selected to represent the symbol for parentheses. In Line 8, I wondered if *furan* and *xiran* also had astronomical referents. Perhaps, I thought, they referred to rising and setting objects in the sky or waxing and waning moons. AA quickly laid my hypothesis to rest in Line 11. This was an ambiguous moment in the interview. I could not understand the speaker's injection about the 'meridian stars' until I replayed the recording, and so I did not follow up on his comments. In fact, further questioning on this point would exceed my knowledge of astronomy, so that the speaker would become for me more than a consultant on linguistic or cultural knowledge, but an instructor of basic science.

In any case, I had another hypothesis about *bil* to investigate at Line 12. For me, the prototypical moon is a silvery disc high in the sky, but the image of the crescent moon as a symbol for Islam passed through my mind as AA described the movement of stars. This hypothesis was confirmed in the final lines of the selection: sensibly enough, it is the crescent moon that offered an appropriate visual metaphor for the parentheses symbols. In English, then, *moon* refers to all the moon's phases and requires adjectives like crescent, new and full to establish the more precise descriptions conveyed in single Somali words. This is the most enjoyable part of linguistic interviews for me, finding out what part of life an idea is located in – astronomy? house-building? travel or trade? – and then imagining how the boundaries and associations of the concepts could differ between languages.

In another section of the same interview, AA explained Somali terminology for angles, based on *xagal*, the word for elbow. This term is interesting mathematically because it embeds the idea of movement. In the United States, students learn angles as stationary measurements first and consider angular change only when they enter a class in trigonometry. But for *xagal*, the possibility of angular movement is implied in the base meaning of the word. AA explained that *xagal fidsan*, 180° angle, calls up the image of something flat, on the plane, that needs to be spread out: as he put it, 'take a carpet and spread it out'.

In contrast, *xagal dabac-san*, angle of depression, refers to something that is not flat, not tight and something that has been released. While I can construct a metaphorical connection here, imagining that an acute angle could be released and stretched out into an obtuse angle, I wonder what situations in daily life would require someone to talk about *dabac-san*:

SS: When would you use that in daily life?

AA: For example, suppose you are a boss and a bully comes and then you are not very strict to him and he's done something wrong, but you ... show him that you are not very serious and that he could do his job but not to, you are not applying to the policy of that ... so we say *dabac-san*. You are not applying to the load, to the policy of that ... you show him something not very stressed ... When you give, not a weak face, but a face that is very cordial, very friendly, very good, very good thing. The bully comes to you, and you are very peaceful. It's being open, flexible, moving.

The ellipses in the transcript represent deletions of repetitive phrases, as AA searched for words to describe a culturally specific concept of leadership. A good leader must address the wrongdoings of others, but must do so in a cordial, flexible and friendly way without showing anger. *Dabac-san* expresses the position that social management rather than retribution is the key characteristic of effective leadership. Unravelling the meaning of a mathematical phrase, then, leads us to core cultural values about authority, control and cooperation.

This selection raises the dilemma of the degree of cultural knowledge that we wish to introduce into students' mathematics learning. The term *xagal* offers opportunities for mathematical thinking about angles in motion, but at the same time, the ordinary meaning of *dabac-san* points towards significant community issues that diverge from students' immediate mathematical needs. One of the consistent problems in Somali immigrant communities is the weakening of elders' authority and of their interaction with the younger generation. Instead of providing guidance to their families, elders in the twin cities often live alone, and frequently feel depressed and ineffective. The police and the judicial system intervene in disputes that elders would traditionally try to resolve (Robillos, 2001), at times using the leadership orientation described as *dabac-san*. A full exploration of the metaphors of mathematical language leads directly into socially vital principles that are no longer fully mathematical.

The example of *dabac-san* offers a point of departure for mathematics teachers to consider how they wish to balance commitment to cultural context with mathematical learning objectives. Mathematics teachers will always establish a boundary between mathematics and the social world that suits their teaching philosophy and goals; many may feel that they have too little cultural knowledge to interact with students on the topic or that the issues of social authority that are embedded in the term *xagal dabac-san* have too little mathematical content to justify their inclusion in a mathematics class. Still, there are ways to bring authentic social content to students without detracting from mathematical learning. Teachers who value strong social contexts for mathematics may prefer to develop non-traditional curricular formats for this type of information: the development of websites, student/community research projects or family involvement activities centred on mathematics meanings.

The Dialogue of Culture and Mathematics

Somali mathematics terms like *bil*, *dabac-san* and *dheeli* establish a historical circuit between culture and mathematics. Cultural objects and practices in Somalia became metaphors for mathematical concepts; today in the United States, learning mathematics terms can re-unite immigrant Somali students with aspects of their cultural history that have been weakened by political violence and interrupted education. Within a mathematics classroom, however, the associations between mathematics concepts and daily life can be viewed as either a resource or as a hindrance to mathematical understanding. In this way, the metaphorical basis of Somali mathematics words opens a dialogue between mathematics and culture that can illustrate multiple viewpoints on educational issues of multilingual classrooms.

A legitimate concern raised by educators familiar with the vocabulary is that the cultural meanings may cloud the precision of mathematical concepts (Pimm, 1987: 92–94). Consider, for example, *saab*, the Somali word for parabola. Culturally, *saab* refers to a basket that covers a ceramic urn used to carry milk or water. A *saab* has a cross-section that approximates a parabola, but the object itself is three-dimensional, and so it might better be associated with the term paraboloid (Favilli, personal communication; Favilli & Jama, 1996). Similarly, the gesture *jibaar*, 'transferred measurement', of laying down a distance several times could be misunderstood as repeated addition rather than as exponentiation.

Even if non-mathematical meanings create the potential for misunderstanding, some scholars understand informal language as a valuable stage on the pathway to technical fluency (Clarkson, this volume; Setati & Adler, 2000) or as a resource for developing mathematical thinking (Leung, 2005; Moschkovich, 2002). Furthermore, learning and using properly the definition for any mathematical term is a difficult process for many students, a road paved with many false tries. At times, the problematic terminology may not be mathematically defined words at all, but rather ordinary words and phrasings in the language of instruction (e.g. Barwell, 2003a). Indeed, Somali mathematics terminology problematizes the entire debate over formal and informal mathematics discourse. Somali mathematics words are at once culturally polysemous and precisely technical; *saab* is just as precise as *parabola* because both terms rely on the same mathematical definition. A student using Somali mathematics vocabulary correctly cannot move from informal to formal speech because this vocabulary is, simultaneously, both formal and informal.

Student perspectives on Somali mathematics terms suggest that the richness of associated cultural meanings stimulates mathematical thought. Most of my Somali algebra students, for example, could name the four arithmetic functions in their home language, and give consistent descriptions of the meaning of *iskudofa*, multiply: 'hitting one thing against another' or 'two things hit together'. Often this was accompanied by a gesture in which their fists were brought together from the upper wrist to the knuckles. One student explained that when you hit two against two, you get four, and went on to suggest that *iskudofa* is like algebraic notation in which $2x$ means multiplication of the coefficient and the variable x. For this student, the adjacency implied by the gesture and meaning of *iskudofa* accurately represented the algebraic convention that adjacency in writing stands for multiplication.

A university student who studied mathematics and science in Somalia described to me his first encounter with the term *fansaar*, function. He explained to me that if a person is riding a horse and offers a ride to a passerby, then the passenger riding on the back of the horse is *fansaar*:

> When I was learning math, I asked myself, 'Why is *fansaar* a function?' Then I thought that a function is something that changes an input into an output, so that there are always two. [SS: Which person is like the output?] The one in back. [SS: That's what I was thinking, too.]

Here, the student reconciled the cultural image of the word *fansaar* with the relationship between the independent and the dependent variable of a function. The dependent rider became the dependent variable.

These examples illustrate students in the process of tracing the metaphorical relationship between ordinary and mathematical meanings. In the language of Lakoff and Núñez (2000), they compel students to explain mappings from concrete to abstract domains of the metaphor. English mathematics terms, by contrast, have, in many cases, lost their etymological metaphors, so that the associated meanings can neither confuse nor support student understandings.

Exploring multiple meanings, in this case, through re-inventing the metaphor that suggested the Somali term in the first place, can help students negotiate mathematical meanings. This is consistent with Moschkovich's (2002) observation that mathematical practices that are taught and valued in many classrooms today move well beyond the acquisition of technical vocabulary. As Leung puts it, 'the teaching of technical vocabulary should be seen as a pedagogic point of departure for exploring concepts, meaning-making and meaning exchanging, not an end point of learning' (Leung, 2005: 134). Student reflections that reconcile mathematical and ordinary meanings of Somali mathematics terms seem to bear this point out.

A Model for Other Languages?

In a famous case study of community-based ethnography, students from a rural school district in the southern United States developed successful, culturally relevant teaching materials (Heath, 1983). Students became deeply engaged in a science project when they began interviewing local farmers; subsequently, the class performed very well on the standard science test. Readers may consider whether mathematical vocabulary could be the focus of similar student and teacher ethnographic studies to enhance mathematics learning for students in their classes. What is the potential for the local development of pedagogically useful, cross-linguistic classroom materials and what problems would a teacher or student ethnographer encounter in developing them? In other words, to what extent could the

Somali project described in this chapter be undertaken for a different home language and a different community?

The project described in this chapter benefits from several unusual factors. Somali is a language with a script that is fairly well known among middle-aged adults. Highly educated Somali men and women work in the educational system and are familiar with the challenges of students learning English as an additional language. The formal mathematics vocabulary is largely defined and, in part, published. Trying to reproduce this project without these community resources could be challenging. For example, presenting mathematical vocabulary in written form will be neither practical nor politically appropriate for some languages that are not widely written. Instead, an audio and video presentation of the data might prove to be an accessible format for students and family members. Audio or video recordings or interactive websites could strengthen students' ability to discuss mathematical principles in their home language and draw the parallels between mathematics and cultural practices that we see among Somali students. Given the dearth of preparation time available to many teachers, student-led research projects might be the most realistic means of gathering information on mathematical terms and ideas represented in their classroom's languages.

Despite significant challenges in trying to recreate the Somali mathematics project in another setting, a parallel project is at least plausible. Creating classroom dialogue on the mathematical words and the mathematical nature of cultural practices would give students a foundation to use in small-group discussions of mathematics in their home language, and it would create a pathway for teachers to introduce culturally relevant materials into their classrooms.

Above all, community-based research responds to the extraordinary diversity of languages that exist in mathematics classrooms around the world. In situations of new immigration, in particular, published teaching materials for many languages may take years to produce. Teachers must have recommendations on creative and timely means of interacting with students who speak a variety of languages, and these approaches must be low cost and flexible enough so that they can be implemented in many situations. Readers of this chapter are invited to consider students' languages in their own classrooms, imagine what methods they would be willing to use and how much culturally relevant information they are willing to include to develop cross-linguistic mathematical learning materials.

Chapter 4
Politics and Practice of Learning Mathematics in Multilingual Classrooms: Lessons from Pakistan

ANJUM HALAI

Pakistan is a linguistically diverse country with over 300 dialects and approximately 57 languages spoken throughout the country's four major provinces (Khan, 2002). While Urdu is the country's national language, it is the primary language of less than 10% of the population (Laporte, 1998). The spoken language in large urban centres in Pakistan is a mixture of Urdu and English known as Minglish, which is common in electronic and print media and is also the language of educated urban youth of relatively upper socio-economic backgrounds (Khan & Khan, 2002).

Pakistan's linguistic diversity is reflected directly or indirectly in the education system. The medium of instruction in government elementary schools (Classes I–VIII, ages 5–13) is generally Urdu, although in the province of Sindh, schools have the option of using Sindhi instead. English is introduced as a second language in Class VI. At secondary levels (Classes IX–X, ages 14–15), the official medium of instruction in the government schools is mainly Urdu. Most of the teachers, however, use their regional language as the language of instruction. Provinces also teach their regional languages as a subject.

Besides the government schools, the private sector is a significant provider of schooling in Pakistan. Schools that use English as the medium of instruction are called English-medium schools. Most private schools are English-medium schools. For parents and students, English is the preferred language of education and is seen as the language of the elite and the ruling class. English-medium schools are found in both urban and rural areas. Students in Pakistan's English-medium schools learn curriculum content and the English language simultaneously, and are expected to become proficient in both (Rahman, 2002). According to Khan (2002) and Haque (1993), almost all Pakistanis would prefer their children to study in English-medium schools because it is seen as a language that opens doors to professional and academic opportunities.

This chapter explores the learning of mathematics in multilingual settings in specific English-medium classrooms in Karachi, Pakistan. I start with a brief history of language in education policy in Pakistan. This historical overview provides the background for a discussion on issues in learning mathematics when the language of instruction is not the first language of the learner. In particular, it sets the scene for the issues related to power and politics of language use in the classroom and raises questions about the role of codeswitching in students' learning of mathematics. The subsequent discussion is illustrated with examples of episodes from English-medium mathematics classrooms. Finally, the chapter ends with a discussion of the implications for policy and practice.

The Mathematics Curriculum

In 2005–2006, the Ministry of Education undertook several reform initiatives in education. This current wave of educational reform in Pakistan has led to the development of a new education policy that aims to provide high-quality education appropriate for a knowledge-based society and to meet the challenges of globalisation. Alongside the new education policy is a curriculum review and new scheme of studies (Ministry of Education, 2006). Improvement in quality, in the case of mathematics education, is seen through an emphasis on problem solving and developing reasoning and logical thinking. Hence, the 2006 curriculum for mathematics is organised in five main standards, of which Standard five is 'reasoning and logical thinking'. Typically, skills like reasoning and logical thinking are developed through a variety of approaches, including a focus on discussion, communication and other discursive approaches to teaching and learning (NCTM, 2000). In Pakistan, this represents a significant shift, since traditionally, mathematics teaching and learning in classrooms in Pakistan is characterised by a focus on rote memorisation of rules and their application to produce a 'right answer' (Halai, 2006, 2007; SPDC, 2003; Warwick & Reimers, 1995).

To make the new national mathematics curriculum relevant to the increasingly global world, proficiency in English is seen as one of the most significant elements of the curriculum reform. Hence, among other things, the new policies include the decision to introduce English as a subject from Class I onwards in all government schools across the country, and to introduce English as a medium of instruction for mathematics, science and computer studies from Class VI onwards by 2011. Other subjects such as social studies and Islamiat (religious studies for Muslims) would continue to be taught in Urdu or in Sindhi in the case of Sindhi-medium schools (Hasan Aly, 2006).

The reform in education described above will be implemented at the level of schools and classrooms. Hence, several issues arise about improving students' learning (a focus of the reform) in classrooms where reasoning and logical thinking in mathematics are emphasised. But, as in the Karachi classrooms described below, any explanation, justification and reasoning is expected to take place in the language of instruction, and not in the languages in which the students are most proficient. Is students' learning of mathematics facilitated or hindered by a change in medium of instruction? Through a microanalysis of episodes from the two multilingual mathematics classrooms described below, this chapter provides insights into this and other related issues.

Language Use in Multilingual Mathematics Classrooms

In multilingual mathematics classrooms like the ones described above, students moved between the language of instruction and their own language in the course of mathematics learning. This movement across languages has been the focus of many studies, mainly in the field of linguistics and socio-linguistics and is often referred to as codeswitching (Boztepe, 2003). While the term codeswitching itself has been used in multifarious ways, for the purpose of this paper, it as an overarching term that covers the phenomenon of moving between two languages. In mathematics education, Adler's (2001) work in multilingual mathematics classrooms has looked at dilemmas emerging from codeswitching, but these are dilemmas for teachers and teaching. In subsequent work, Setati and Adler (2000) have proposed that codeswitching is a practice that enables learners to harness their main language as a learning resource. Investigating why students codeswitch in multilingual mathematics classrooms, Clarkson maintains that swapping languages makes it easier for students to understand the problem when it is perceived to be difficult and that sharing a language sets an emotional tone that is conducive to group thinking (Clarkson, 2005: 37). In these studies, there is a recognition that 'language switching' and 'language swapping' are something that occurs in many multilingual classrooms and that as a mechanism for learning, codeswitching has almost become a taken-for-granted 'good thing'. Is this always the case? The phenomenon of codeswitching as a 'good thing' needs to be problematised, since we know relatively little about students' experiences of multilingual classrooms (Sierpinska, 2002).

There has been relatively little research on students' learning in multilingual mathematics classrooms. However, the emerging knowledge base suggests that a socio-political lens, rather than a psychological lens, is needed to get insights into why students swap languages as they engage with mathematics. For example, Clarkson elaborates on students' reasons for codeswitching as

> the influence of important others or situations, the difficulty of the problem itself, or just because they feel like it. But within their solution processes understanding the ways in which these students use their

languages and why has the potential of providing rich insights for students themselves as well as teachers, curriculum developers, and researchers. (Clarkson, 2005: 39)

The 'influence of important others' suggests that reasons for codeswitching are not just rooted in concerns of cognition and communication; rather, they are also linked to issues of power and status. Gorgorió and Planas (2005) raise further issues regarding the politics of language use in mathematics classrooms. Based on their research with immigrant children in Spain, they maintain that in the course of learning, students interpret classroom processes based on their cultural and social backgrounds. While this would be the case with all learners, in multilingual and multiethnic classrooms, there is positioning of the immigrant context as 'different' from the common or normal context familiar to the teacher. This positioning raises issues of power and status because the immigrant context is subordinated in comparison to the dominant context. Cummins (2000b) maintains that in order to promote learning and achievement for learners for whom the language of instruction is not their first language, it is not enough to focus on the language of instruction. There is a need to examine the hidden curriculum being communicated to students through that instruction. It is through challenging the coercive power structures that position one language in dominance over others that an ethos of empowerment would be created. He supports the idea that educators who strive to create educational contexts within which culturally diverse students develop a sense of empowerment through bilingual education are, of necessity, challenging societal structure and power (Cummins, 2000b: 247–258). Drawing from his work with two British students of Pakistani origin, for example, Barwell illustrates how the students were able to support their learning of mathematics through empowering interactions that allowed them to draw on their shared cultural and linguistic experiences (Barwell, 2005b: 463).

In post-colonial contexts like Pakistan, socio-political elements influencing the switching of languages are rooted in the relative positioning of the language of instruction and their own language. According to Rahman

> In Pakistan as we have seen, English is not so much imposed as rationed. It confers much prestige and has high utilitarian value, giving privileged access to the international and the most powerful national salariat groups, and is greatly in demand. (Rahman, 2002: 533)

The socio-political dimension of language use in classrooms, in particular the policy of using English as a medium of instruction, is also found in other post-colonial contexts. For example, Setati (2002) traces the controversial history of official language policy in South Africa. She maintains that 'language-use in a multilingual educational context . . . is as much, if not more, a function of politics as it is of communication and thinking'

(Setati, 2002: 6). Understanding the politics of language use in classroom interactions could help us understand why students move between languages in multilingual mathematics classrooms. This key could open the door to understanding aspects of learning that a psychological approach would not allow.

To explore how some of the above issues appear in mathematics classrooms in Pakistan, this chapter draws on findings from a research project (Halai, 2001) that examined the role of social interaction in students' learning of mathematics as they worked in small groups in two classrooms in Karachi. Over a period of one academic year, lessons were observed and recorded on video tape. The observations focused on social interaction within groups of students as they engaged with mathematics in the course of the lesson. In addition to the observations, stimulated recall interviews with students and post-observation conversations with the teachers were conducted.

Two Mathematics Classrooms in Karachi

The two mathematics classrooms discussed in this chapter were both from an English-medium school in Karachi. The two classrooms, one Class VI (10–11 years old) and the other Class VII (11–12 years old), were multilingual, that is, the language of instruction was not the first, often not even the second language of the learners. Students came from Gujrati-, Katchi-, Sindhi- and Urdu-speaking backgrounds. However, all of the students were fluent in the use of Urdu. Neither teachers nor students were fluent in the use of English. The two teachers were graduates of an innovative teacher education programme at a local university. As a result, these teachers had initiated a change in their classrooms by introducing group work and open-ended mathematics tasks purported to be closely related to students' everyday experiences. This is the kind of approach expected in the new curriculum.

Teaching in the two classrooms was almost entirely in English and the textbooks used and the tasks set in the class were also in English. The teachers' formal introduction to the lesson, worksheets and instructional materials were never given in Urdu. But during group work, students spoke in a mixture of Urdu and English and the teachers also used Urdu when they discussed the work with the various groups. In reporting the group work to the whole class, the students spoke in English with a smattering of Urdu. However, the mathematical terms used in this mixture were invariably in English because these terms came from the textbook, which was in English. Similarly, the groups, when presenting their work to the whole class, always spoke in English. At times, students who had difficulty in the use of English *took permission* from the teacher to report entirely in Urdu. All these observations suggest that there was a tension in the position that was enjoyed by English in the mathematics classrooms relative to the position enjoyed by Urdu. While English appeared to enjoy the patronage of teachers' and authorities such as the principal and the education policy makers, Urdu was seen as something to be used away from the central scene in the classroom, something not to be acknowledged officially.

Episodes taken from the two classrooms are presented to illustrate some of the issues that arose for learning mathematics in multilingual classrooms. These issues are related to the role and purpose of codeswitching by students, understanding and abstracting the mathematical intention from everyday language used in mathematics texts and tasks and the politics of language use in classrooms.

Lesson episode 1

In a lesson on ratios, the students in Class VII (11–12 years old) worked on the task named 'Anisa's Drink'. The task involved ratio comparisons when the drink, made from mixing *Rooh Afza* and water, was prepared by Anisa. *Rooh Afza* is a local drink prepared by mixing *Rooh Afza* liquid and

ANISA'S DRINK

3. Here are three Rooh Afza and water drinks.

Which of these sentences are true and which are not? Give reasons for your answers.

a   The Rooh Afza drink in C is the strongest.
b   The drink in C is Rooh Afza and water in the ratio 6 : 9.
c   The drink in B is Rooh Afza and water in the ratio 2 : 3
d   All the drinks have the same strength.
e   Drinks A is weaker than drink B because it is Rooh Afza and water in the ratio 2: 3 and the ratio for B is 4 : 5
f   A drink made with Rooh Afza water in the ratio 10 : 15 will be stronger than the drink in A.
g   If you used 10 times more Rooh Afza and 10 times more water in A the Rooh Afza drink would be 10 times stronger.

4. A Rooh Afza drink is made from Rooh Afza water in the ratio 1 : 5 which of these mixtures of Rooh Afza to water have the same strength?

    (A) 2 : 10   (B) 2 : 7   (C) 10 : 50   (D) 2 : 6

Figure 4.1 The Rooh Afza problem

water in varying quantities. In the excerpt shown below, students Faizullah, Mansoor and Saleem are working from a worksheet (see Figure 4.1) using a combination of Urdu and English. In the transcripts, I include translation in English, in italics, wherever Urdu has been used.

0   **Saleem:**   (reads) Which of these sentences are true and which are not? Give reasons for your answers.
(Some mumbles and a pause)

1   **Mansoor:**   (reads question 3, statement a) 'The Rooh Afza drink in C is the strongest'.

2   **Saleem:**   Ab dekho C ka strongest kyun hay?
Now look why is C strongest?

3   **Mansoor:**   Haan strongest hay.
Yes is strongest.

4   **Saleem:**   Kyoon.
Why?

5   **Mansoor:**   Iss mein ziada hay naa.
It has more that is why

6   **Saleem**   Pahley sab ke ratio likh laitay hain.
Let us first write the ratios of all (Saleem writes all ratios)

7   **Saleem:**   Ab dekho (points to the writing) konsay (ratios) barabar hain?
Now look (points to the writing) which (ratios) are equal?

8   **Mansoor:**   Koi barabar nahi hai
None are equal

9   **Faizullah:**   Saab saay strongest kya hai?
Which is the strongest of all?

10   **Mansoor:**   C hai. C is the strongest. Vo strongest hai. Iss mein ziada hai na
C is. C is the strongest. It is the strongest it has more that is why
(Slight pause)

11   **Faizullah:**   Drink in C is the strongest. Haan na tau hai na
Drink in C is the strongest. Yes so it is

12   **Mansoor:**   Ziada hai na. Iss (drink A) mein dau hai tau iss mein kaisa ho ga. Yeh (drink C) ziada log piyain gaay tau vo strong ho ga
It (drink C) has more. It (drink A) has two so how can this be? This (drink C) more people will drink so this will be strong
(Laughter)

13   **Saleem:**   Yeh (drink B) bhi tau char hain aur iss mein panch cup hain
This (drink B) is also four, and it has five jugs

14	**Faizullah:**	(laughs)
15	**Saleem:**	Haan botlain barabar hain. Magar iss mein (refers to drink B) pani kam hai, ziada tez tau yeh hi hua na (compares B and C)
		Yes the bottles are the same but in this (refers to drink B) water is less so this will be more strong.
16	**Mansoor:**	Yeh kyun hua?
		Why is this so?
17	**Faizullah:**	Kya hua bhai?
		Why is this so brother?
18	**Saleem:**	Char botlain iss mein bhi.
		Four bottles in this one (refers to B)
19	**Mansoor:**	Haan
		Yes
20	**Saleem:**	Iss mein bhi char botlain. Iss mein bas 8 cup hain iss mein ...
		In this (refers to C) also four bottles. In this (refers to C) there are eight cups ...
21	**Mansoor:**	In this (refers to B) there are five
22	**Saleem:**	Five jugs so why are you looking at five jugs?
23	**Mansoor:**	Okay okay
24	**Faizullah:**	In this (drink B) there is more Rooh Afza silly, almost one bottle (laughs and looks at Mansoor).
25	**Saleem:**	C is not strongest
26	**Faizullah:**	Yeah

In Urdu, the word '*ziada*' is mainly used to mean 'more of something'. Moreover, in Urdu, degrees of comparison are shown in one of two ways. In one, the drink could have been *tez* drink (strong drink), *tez taar* drink (stronger drink) or *tez tareen* drink (strongest drink). In this case, the word *tez* stands for strong. Alternatively, the degree of comparison could have been shown by pre-fixing the word *ziada* (more) to the attribute, which is supposed to be compared. Hence a *tez* drink (a strong drink) would be *ziada tez* drink (more strong drink). The issue of students' usage of the word *ziada* becomes compounded because the students are not translating the word 'strong drink'; rather, they are using a mixture of Urdu and English.

There is a tension in the way the word '*ziada*' was being used in this excerpt. The one usage was 'more in volume', and the other was 'more in relative concentration'. Mansoor started by stating the *Rooh Afza* drink in C was the strongest (Line 1). His reason (in Lines 5 and 10) that drink C was strongest was that 'the drink C has more'. More of what is not clear: water or *Rooh Afza*? It appears that Mansoor was using the word '*ziada*' in the sense of more in volume because in Line 12 he explains '*ziada log piyain gaay'*, that is, more people will drink, suggesting that the greater volume of the drink will be enough for a greater number of people. An interpretation is that Mansoor regarded the word 'strong' to mean more, so that his interpretation of 'strongest drink' was bounded by the context of the problem task, that is, that 'strongest drink means more drink'. Saleem's explanation (Line 15) uses the word *ziada* in the sense of ratio. Saleem appeared to recognise that there was a particular usage of this phrase. So, he tried to enable Mansoor and Faizullah to interpret the phrase in the sense of its usage in the problem task, that is most *Rooh Afza* drink per unit quantity of water. Saleem (Lines 13 and 15) encouraged Faizullah and Mansoor to focus on the *Rooh Afza* and water in Drinks B and C. These focusing statements by Saleem appear to have led Faizullah to change his thinking. He questioned Saleem's statement in Line 15 by asking the question in Line 17. However, from his muttering tone, it seemed that Faizullah's question was directed to no-one in particular; rather, he asked this question of himself. The question seemed more for the purpose of Faizullah's own reflection. Saleem continued with his conversation with them until Faizullah apparently suddenly declared in Line 24, that, 'In this (Drink B) there is more Rooh Afza silly, almost one bottle'. Hence, in this case, insight into ratio comparison seemed to have occurred for Faizullah when he decided that Drink B was the strongest because it had *almost one bottle* of *Rooh Afza* for a jug of water.

Lesson episode 2

Subsequent to the above task, I asked students to identify, from four different combinations, mixtures of *Rooh Afza* and water that have the same strength and asked them to provide reasons for their answers. This subsequent work on ratio tasks showed that the students, including Saleem, used additive, rather than multiplicative reasoning. For example, in the data excerpt below, students are considering the following task:

> You can see the height of Mr Short measured with 6 paper clips and 4 large buttons. Mr Short has a friend Mr Tall. When we measure their heights with large buttons, Mr Tall is 6 large buttons in height. How many paperclips are needed for Mr Tall's height? Explain. (Karplus et al., 1979)

1	**AH:**	Now I am asking Faizullah the reasoning. Saleem, I will come to you. Let me ask Faizullah first, the reasoning, because Mansoor has given his reason. Yes?
2	**Faizullah:**	6 and 4 and 8 …
3	**AH:**	Yes
4	**Faizullah:**	… 6 and 8 oh 6 and 8 because err 4 and 6, 4 plus 2 is 6 and 6 plus 2 is 8.

5	**AH:**	Okay what do you think, Saleem?
6	**Saleem:**	Teacher, 8 paper clips, because teacher, we are adding 2.

The incorrect strategy used by the students was to focus on the difference. Although the use of additive reasoning is consistent with research findings elsewhere (Hart, 1981; Noelting, 1980; Sowder *et al.*, 1998), I was surprised because Saleem and Faizullah seemed to have used multiplicative reasoning previously in the task on Anisa's Drink.

My findings regarding students' learning of ratio confirm the claim by Sowder *et al.*, that

> There is some consensus that additive reasoning develops quite naturally and intuitively through encounters with many situations that are primarily additive in nature. Multiplicative reasoning does not develop so naturally; schooling is required to develop a deep understanding of multiplicative situations and appropriate responses in these situations. (Sowder *et al.*, 1998: 120–129)

Hence, one interpretation of the students' inability to perform on ratio tasks could be that the concept of ratio and quantitative reasoning is inherently complex and, therefore, difficult to learn. Indeed, students' varied performance on the ratio tasks suggests that they were at an early stage of development in their understanding of quantitative reasoning. Another interpretation is that language patterns and discursive practices in the classroom are supposed to enable students to abstract the mathematical concepts and relationships (Sfard *et al.*, 1998). But, in this setting, language use was itself problematic so that students' negotiation of the mathematical meaning of ratios and proportions came into conflict with structures of the language of instruction and with the mathematical expectations embedded in the everyday context. I believe that the second interpretation bears some weight because there were similar examples from subsequent lessons in the same class. For example, in a lesson on proportions, the teacher asked the students to find the 'fair share' in the profit when two friends had set up a business, investing capital in a ratio of 2:3. Mansoor translated the phrase 'fair share' as *barabar hisaay*, that is equal shares. Hence, their calculation of shares in profit was equal and not proportional, which would be the mathematical essence of a fair share.

Lesson episode 3

In this lesson on equations, the students Maheen, Naima, Samina and Shabnum, from Class VI were asked to 'find equations for the following problem statements and solve them':

Q1. Think of a number and subtract 6 from it, the resulting number is 10. Find the number

Q2. Sara will be 28 years old after 9 years. Find her present age.

Q3. Sum of two numbers is 115. If one them is 63, find the other.
Q4. After giving 18 marbles to Anisa, Asad has 45 left. How many did Asad have originally?

In the excerpt below, students are working at Q2.

| 1 | **Shabnum:** | Sara will be 28 years old after 9 years. *Will be next* (emphasises using body movements). |
| 2 | **Maheen:** | 'Will', nine years ke baad
Will after nine years |
| 3 | **Samina:** | Iska matlab yeh hai ke Sara 28 years ki hai
It means that Sara is 28 years old |
| 4 | **Shabnum:** | Nahi 28 years *ki hogi* because after nine years
No, will be 28 year because after nine years |
| 5 | **Maheen:** | Because yehan 'will' (hai) 'will' means future
Because there is 'will' here. Will means future |
| 6 | **Shabnum:** | hogi nahi nahi vo hogi nine years ke bad
Will be, no, no she will be after nine years |
| 7 | **Maheen:** | hogi aise batao ke yehan will hai will means future tense
Will be, tell her that here there is 'will' (and) will means future tense |
| 8 | **Naima:** | han han bhai chalo
Yes yes move on |

In the above extract, it is evident that understanding how the word 'will' is used is crucial to the students' successfully doing the task. Knowing that 'will' is future tense has a major implication for how the problem statement is converted into a mathematical equation and then a solution is sought. This episode indicates that proficiency in the language of instruction is necessary to understand the statement of the mathematics tasks because abstracting the mathematics appeared to be contingent upon understanding the language in which it was encoded. It is a significant issue that has implications for the policy and practice of language use in classrooms. For example, if the policy requires students to learn mathematics alongside the language of instruction, as is the case in the proposed reform in mathematics education in Pakistan, the quality of learning will necessarily be compromised.

There were other similar examples of issues that students faced in interpreting mathematical tasks that require an understanding of the grammar and usage of words in a second or third language. For example, in a lesson on mode, the teacher constructed a task where the phrase 'most frequently appearing' was used in various everyday situations such as the following:

- The number of hours spent by each of six students in reading a book are 6, 6, 8, 11, 14, 21, Which number is appearing most frequently?
- The number of people living in each house on Garden Street are 2, 2, 3, 3, 4, 6, 8. Which number is appearing most?

- Saeed Anwar's runs scores at Australia are 72, 80, 86, 93 and 94 in five one-day matches. What is the most frequently appearing value?

The mathematical examples presented by the teacher could have been misleading, since in the second item there are two numbers that appear twice each, while the third item has no mode. Moreover, the wording of the third item mixes up the frequency of data items and the value of data items. Ensuing class discussion showed that students' focus on the values of the data item led to some confusion about the meaning of mode, that is, confusion between the most frequently appearing data item or the biggest value. Hence, Naima responded as follows:

> **Naima:** (reads the next question) 'Saeed Anwar runs scores at Australia are 72, 80, 86, 93, and 94 in one day matches. What is the most frequently appearing value?' The answer is 94 because the biggest value in 72, 80, 86, and 94 is 94. That's why my answer is 94.

On the one hand, this confusion is a mathematical issue because the value of a data item is different from its frequency of occurrence. On the other hand, it is also a language issue because the teacher, being knowledgeable in mathematics, could be expected to know the difference in value and frequency of occurrence of data items. But the same could not be said about his knowledge and vocabulary of the English language. Although the focus of this paper is on students' learning, issues of teachers' proficiency in the language of instruction are also worth taking into account. As is evident from the classroom data shared above, teachers' proficiency in the language of instruction is important for them to be able to communicate clearly and accurately the mathematical intent embedded in the mathematics tasks and texts.

Politics and Practice of Learning Mathematics in a Multilingual Classroom

There are three points that can be made arising from the preceding episodes. First, the practices that students engaged in as they undertook mathematics showed movement between the language of instruction and their own language. This movement across languages involved a demonstrated need on the part of the learners to understand the language structures, grammar and vocabulary of the language of instruction. It also involved translation, which is a nuanced and complex process. Due to these complexities, questions arise about the role of codeswitching in aiding the process of learning mathematics. Second, the discourse of mathematics classrooms, where the mathematics texts employ 'everyday phrases' in the language of instruction, requires understanding the mathematical intention as well as understanding the language of instruction.

Third, the politics of language emerged as a significant element in the classroom dynamic, thereby suggesting that the role of language in students' learning should be seen as beyond communication and cognition. These points are now discussed in more detail.

Role of codeswitching in learning mathematics in multilingual classrooms

As students worked at mathematical tasks it appeared that their understanding of the statement of the problem task required interpretation on at least two levels. One level was understanding the language involved and the other was understanding the mathematics involved. It is reasonable to assume that the learners in any classroom would need to understand the language of instruction before they make sense of the mathematics encoded in that language. However, in a multilingual classroom, interpreting the language of instruction posed an additional challenge because the language of instruction, and hence the problem, was stated in English, while the students switched code and used Urdu in thinking aloud about the problem, doing work on it and explaining it to their peers. This observation of students reverting to Urdu suggests that they were more comfortable sharing their thinking in Urdu and that the use of English to do this might be problematic for them. An interpretation could be that the group interactions, being in Urdu, were aiding students in their effort to learn mathematics meaningfully. Typically, similar interpretations have led the role of codeswitching to be seen as facilitating learning (Clarkson, 2005; Setati & Adler, 2000). However, the evidence from the classroom episodes discussed above suggests that codeswitching may not always facilitate learning. For example, in Episode 1, the use of the word 'stronger' had to be understood as 'more in relative comparison'. And there are specific ways of showing degrees of comparison in Urdu, which are different from the ways they are shown in English. Students' usage of the word 'stronger' as *ziada* in Urdu does not reflect the second degree of comparison. It was this mathematically appropriate usage of stronger that was key to students abstracting the essence of ratio comparison. Similarly, it was a proportional share that would be regarded as a fair share in the context of a mathematics lesson on proportions. Likewise, in Episode 3, it was the understanding of 'will' as future tense that would enable students to do the mathematics correctly. Hence, issues of grammar, vocabulary and language structures have added significance in multilingual mathematics classrooms. These observations raise questions about the 'facilitative role' of codeswitching. It also raises questions about the change in language in education policy in Pakistan: should the introduction of English as a medium instruction not be delayed until students are more proficient in the use of English language?

Everyday life discourse of mathematics texts

An issue that emerged for learners as they worked through mathematics tasks in the multilingual classroom was their apparent need to understand the specific use of everyday life phrases in the mathematics texts and in teachers' instructions for the mathematics tasks. The mathematics texts and tasks used in the classroom episodes were discursive in nature and valued students' everyday experiences, so that it is the mathematical values and perspectives embedded in everyday language that the students were expected to abstract. There are several layers of understanding before students can abstract the mathematical intent of the everyday language. They have to understand the everyday use of the language of instruction and then understand the mathematical intention of this everyday use. In this case, it meant, among other things, understanding the use and purpose of those everyday words in English that the teacher used to 'facilitate' students' learning. Classroom interactions showed that students failed to understand the mathematical purpose of everyday words in English and could not link them to mathematics concepts or relationships. For example, 'stronger' to mean more in proportional comparison and 'fair share' to mean proportional share. Although this could also be an issue in monolingual mathematics classrooms, in multilingual classrooms it raises questions about the extent to which students' understanding of the language of instruction was at play. These observations point towards the arguments presented by Cummins (2000b) that in order to understand the factors affecting students' participation and learning of mathematics in multilingual classrooms, it is important to look beyond the language of instruction and examine the hidden curriculum being communicated to the students. In other words, issues of learning mathematics in multilingual classrooms are rooted in the politics and culture of the classroom.

The politics of language use in classrooms

The classroom episodes shown above illustrate how the students were negotiating meanings for mathematical concepts and ideas that were embedded in the everyday daily-life contexts and coded in the language of instruction. The students invariably used Minglish and reverted to codeswitching when they engaged with mathematical reasoning and addressing the mathematics problems. In other words, codeswitching was an integral element of the classroom interactions and appeared to have the teachers' tacit approval. However, codeswitching occurred when the students were working in groups or not in front of the whole class. Urdu was seen as something to be used away from the central scene in the classroom, something not to be acknowledged officially. On the occasions that students used Urdu in front of the class, they sought permission from the teacher. These observations suggest that classroom interactions reflect and/or reinforce the broader societal patterns of coercive relations of power between dominant and subordinate languages. Clearly, if codeswitching is seen as facilitating learning, it would need to be recognised as such and be given a central place in the learning processes in multilingual classrooms. Of course this means challenging the status quo with respect to the positions of languages in society. Furthermore, as explained earlier in the chapter, using a mix of English and Urdu is a behaviour portrayed by the social elite, as used, for example, by the young and upcoming in the media. Was Mansoor, in Episode 1, using the word 'stronger' because he did not know of a suitable word in English? Or was it a way of identifying with the social elite?

Implications and Recommendations

From the discussion so far, it can be concluded that learning mathematics in multilingual classrooms requires that students understand the language of instruction to be able to understand mathematical ideas and concepts. Moreover, in the case of classrooms where teachers try to facilitate learning by developing curriculum materials based on students' everyday experiences and making use of everyday phrases, this understanding entails gleaning the mathematical intention of the everyday phrases while learning the language of instruction. Situated in the context of the new Pakistan national curriculum, these conclusions and observations have strong implications for policy and practice. As stated earlier the new national curriculum in mathematics has five standards one of which is 'Reasoning and Logical Thinking' (Ministry of Education, 2006). Reasoning and logical thinking are processes that are supposed to promote mathematics learning with understanding as opposed to learning by rote (NCTM, 2000). A direct implication of promoting reasoning and logical thinking for mathematics teachers and teacher educators is to enable students to become proficient in communicating mathematical ideas and relationships through various oral and written approaches. But, the above analysis suggests that factors, processes and developments that apply to mathematics learners in multilingual classrooms would be different from those where learners use their first language to learn mathematics. For example, the prevalence of codeswitching in multilingual classrooms suggests that there is a need for teachers, teacher educators and policy makers to look into ways of maximising the potential of codeswitching through appropriate policies, teaching practice or curriculum materials. Furthermore, to avoid a potentially confusing use of the language of instruction, the students and teachers need to be proficient in the use of language of instruction. A recommendation for policy and practice is that the use of a language as a medium of instruction (English in this case) should be delayed until students are proficient enough to understand the mathematics that is encoded in that language of instruction. Likewise, policy on core competencies required for becoming a mathematics teacher in a multilingual mathematics classroom should include proficiency in the language of instruction.

More significantly, the observations and discussions show that language use and learning mathematics is not simply an issue of cognition and communication. Rather, the status of the dominant language, in this case English, sends messages to the students that *their* language, in this case Urdu, does not have official patronage in the classroom and potentially does not have anything to contribute to the processes of teaching and learning. The socio-political dynamics of languages in multilingual classrooms and patterns in students' interaction raise questions that have strong implications for supporting student learning and that potentially have their resolution in the politics of language in the classroom.

Codeswitching is a reality in multilingual mathematics classrooms in Pakistan and appears to be a resource for enabling mathematics learning for students for whom the language of instruction is not the first language. However, for clear recommendations and guidelines for policy and practice more research is required into understanding why learners move across languages and how the process facilitates or hinders mathematics learning.

Chapter 5
Mathematical Word Problems and Bilingual Learners in England

RICHARD BARWELL

In England, as in many other parts of the world, a significant number of school students face a double challenge: they must learn mathematics even as they are learning the language used in the classroom for the teaching and learning of mathematics. Cynthia, for example, is from Hong Kong. Cynthia went to school in Hong Kong and speaks Cantonese. She is nine years old. Eighteen months before I met her as part of a research project (described below), she and her family moved to England. She now goes to a primary school where almost everything happens in English. She is learning this new language and making good progress: at this point the school has assessed her proficiency as 'becoming familiar with English'. Cynthia, then, must participate in and make sense of the daily mathematics lesson, while at the same time continuing to learn English.

During Year 5 (9–10 years old), Cynthia took a mathematics test. One of the questions is the arithmetic word problem shown in Figure 5.1. In the space following the question, Cynthia added £1.70 to 65p, giving £2.35 as her answer. It is not difficult to see how Cynthia may have struggled to interpret the word problem in the way that the writers of the test intended. Who, after all, is Mrs Patel? What is she doing? We might deduce that she is in a shop of some kind, but this is not made explicit. Why is she buying so many milkshakes? What flavour are they? What kinds of sandwiches does she buy? Who are they all for? Although the problem is simply

---

Mrs Patel buys **4 milkshakes** costing **65p each** and **3 sandwiches** costing £1.70 each.

Work out the **total cost**.

Figure 5.1 An arithmetic word problem (QCA, 1998)

worded, in other ways, it is linguistically rather complex. There is much that is implicit and much that requires a familiarity with this kind of problem in order to make suitable interpretations. These kinds of linguistic demands are, in the case of word problems, fairly specific to school mathematics and are likely to take considerable time to become part of students' English language proficiency (e.g. Thomas & Collier, 1997). How, then, do bilingual learners like Cynthia interpret word problems? How do they work with word problems as part of doing mathematics? And how can teachers support bilingual learners who are learning the classroom language to engage with what word problems require?

In this chapter, I address some of the issues raised above, drawing on my research in Cynthia's school. The main data for this research (which is described in more detail below) consisted of recordings of bilingual learners, including Cynthia, working with a partner on a word problem task. Through extensive analysis of these data (e.g. Barwell, 2003a, 2005a, 2005b, 2005c), I have become aware of some of the ways in which the task supported the bilingual participants to make sense of word problems. In this chapter, then, I give an overview of this work, highlighting three aspects of the students' sense making in particular. These ideas are illustrated with extracts from recordings of Cynthia's work. Before proceeding, however, I need to provide some background, first about word problems and then about the context of the research.

Word Problems and Bilingual Learners

Word problems, like the one shown above, are recognised as presenting difficulties for many students (for a thorough account, see Verschaffel *et al.*, 2000). The rationale for using word problems (apart from tradition) is that they require students to apply mathematics to 'real life' situations. Given their rather stylised nature, the extent to which they do this is perhaps questionable. Nevertheless, researchers continue to be baffled by many students' apparent inattention to the real life setting of a given problem (e.g. Lave, 1992; Verschaffel *et al.*, 2000). In the case of Cynthia's response to the word problem shown above, for example, her addition of £1.70 to 65p makes little sense within the shopping context in which the problem is implicitly embedded, although we should be aware that we do not know how Cynthia interpreted that context.

Word problems are also the mathematics task most investigated by researchers interested in bilingual learners of mathematics (e.g. Adetula, 1989, 1990; Bernardo, 1999; Clarkson, 1983, 1991, 1992; Clarkson & Galbraith, 1992; Mendes, 2007; Mestre, 1986, 1988; Secada, 1991). This interest perhaps reflects the position of word problems at the intersection of linguistic and mathematical thinking, requiring students simultaneously to make sense of a highly specific linguistic form, draw on their social and cultural experience to understand the scenario portrayed in the problem and recognise and solve an implied mathematical problem. In the above problem, for example, there are social and cultural assumptions about shopping, diet and people's names. The reader is also expected to realise that what the question actually requires them to do is not to think about Mrs Patel or her milkshakes, but to perform a combination of addition and multiplication calculations and write down the outcome (see, e.g. Cooper, 1994; Cooper & Dunne, 2000). Not surprisingly, word problems have the potential to be particularly difficult for bilingual students like Cynthia.

Several studies have compared bilingual students' performance on word problems, either with monolingual students (e.g. Clarkson, 1983, 1991, 1992; Clarkson & Galbraith, 1992; Secada, 1991) or when using different languages (e.g. Adetula, 1989, 1990). This kind of research is beset by the complexities inherent in investigating mathematical attainment across languages. It is more or less impossible to design a study of this kind that can separate language effects from mathematical effects, particularly in the case of word problems. This complexity is apparent in a study conducted in the United States by Mestre (1986), who compared groups of monolingual and bilingual Hispanic university engineering students. He tested them in reading in English, algebra and on a set of word problems. He found that the Hispanic students worked more slowly on the reading tests than the monolingual students. There was little difference between the two groups on the algebra test, but on the word problem test the Hispanic students were both slower and less accurate than the monolingual students. The results of the algebra test suggest that this difference is not necessarily due to Hispanic students being less proficient at mathematics. Furthermore, the level of vocabulary in the word problem questions was suitable for the bilingual students' level of English proficiency. This suggests that there are other interpretative demands on students apart from vocabulary. Mestre concludes, 'knowing the vocabulary in a word problem is no guarantee that the mathematical relationships ... will be appropriately interpreted [by non-native speakers]' (p. 399). Certainly, being able to read a problem is not the same as making sense of it in the way the writer of the question intended. Lower scores than those achieved by monolingual students may be due to the reading and interpretative demands of the tests, rather than students' proficiency in mathematics. In Cynthia's case, therefore, it may be that linguistic factors explain Cynthia's difficulty with the above question. Cynthia is quite capable of reading the test question shown above. She can also make sense of it in some way; it is meaningful to her. What is required by the examiner, however, is that Cynthia works out what she is *meant* to do, what sense she is *supposed* to make.

Word problems, then, make high linguistic demands on students. It would be useful, therefore, to examine the linguistic features of word problems in more detail. Gerofsky (1996) has conducted such an examination, analysing word problems as a kind of genre. A genre is a type of text having some kind of identifiable structure or pattern. Written genres include, for example, novels, letters, poems and school reports. School mathematics includes various written genres, including those produced by students, such as their solutions to problems or longer investigative pieces of work (see Morgan, 1998). Students also encounter written mathematical genres such as mathematical questions or problems, which are generally imported into the classroom via textbooks or assessment documents.

In her analysis, Gerofsky (1996) in effect identified features of word problems that make them recognisable *as* word problems. I will mention two of these features in particular. First, Gerofsky identified a three-part structure, consisting of

- a 'set-up' to establish a scenario or minimal story-line;
- a number of items of information;
- a or some question(s) (p. 37).

In the case of the example shown in Figure 5.1, the set-up consists of little more than 'Mrs Patel buys'. The problem continues by providing several items of information: four milkshakes, 65p each, three sandwiches, £1.70 each. The problem concludes with a 'question', in this case in the form of an instruction: 'work out'. The second feature identified by Gerofsky that I want to mention is that the scenarios that contextualise word problems have only a general bearing on the information components of the problem (Gerofsky, 1996: 41). Such information is, therefore, interchangeable and whole classes of problem are possible within the same scenario. So, for example, Mrs Patel could buy five milkshakes costing 50p each or any number of milkshakes costing more or less anything.

Gerofsky also highlights some of the specific linguistic features of the word problem genre. These features include the common use of present continuous aspect (Mrs Patel *buys*, rather than *bought*) and the unusual way in which references to people or objects effectively bring these things into being (when we read 'Mrs Patel' we must assume the existence of such a person, of which we previously had no knowledge). Of course, not every word problem will conform to this set of features, but most word problems will have most of them, for if they deviate too much from the generic structure, they cease to be word problems at all.

The generic features of word problems are likely to be particularly significant in the case of bilingual learners. One of the functions of genre is to support the reader in making sense of the text. Advertisements, for example, are usually recognisable as advertisements, and this recognition influences how we interpret what they say. Similarly, word problems are more understandable when we are familiar, even tacitly, with their generic structure. In the world of word problems, for example, the 'real world' motives and thoughts of named characters are not relevant (unlike, say, in stories or cartoons). Thus, it is not relevant to consider why Mrs Patel is buying milkshakes, let alone why she buys four, why they are so cheap or what flavours she asks for. Learning how to solve word problems is as much a case of learning the 'rules' of word problems, as of being able to do arithmetic or read the English language. Word problems, then, are highly complex linguistic forms and bilingual learners may take some time to develop awareness of some of their subtleties.

Writing Word Problems: Researching Bilingual Learners in the Mathematics Classroom

The work I discuss in this chapter is based on analyses of the discussions that took place between pairs, or occasionally groups of three students, including at least one bilingual student, working on a particular task (described in Barwell, 2003a, 2005a, 2005b, 2005c). The students were asked to jointly write and solve word problems of their own. More specifically, I asked them to write word problems 'about' addition or division. I audio-recorded the students as they worked. How did I come to make these recordings? To explain, I will say a little about the school, the class and how I did the research.

The research was conducted in a medium-sized primary school in a medium-sized city in England, and was designed to investigate how students learning English as an additional language participate in mathematics classes. I spent two years visiting one of the teachers in the school. In each of those years, she taught the Year 5 class of around 24 students, of which six in each year were bilingual learners. These students came from a variety of backgrounds in East Asia, East Africa and South Asia. Some, like Cynthia, had recently moved to England. One student, Safia, for example, had moved from Somalia to the Netherlands where she went to school, before moving again, first to Wales, and then to England. Others had been born in England and grown up in households where two or three languages were used. Two students, Farida and Parveen, were born in England and attended the school, but they had also attended school in Pakistan on extended visits with their families. All of the students involved in the research had been identified by the school as learners of English as an additional language and most of them had been evaluated as working at a level known as 'becoming familiar with English'. This level of proficiency was known as 'stage 2', where 'stage 1' was referred to as 'new to English'.

Mathematics was taught every morning in Year 5, after the mid-morning break. As recommended by the prevailing national guidelines for teaching mathematics (DfEE, 1999), each lesson began with a period of teacher-led whole-class work, followed by small group work. Lessons usually ended with whole-class discussion of the morning's work. The teacher had identified word problems as an area of the mathematics curriculum with which her students often struggled, particularly the bilingual students. It was apparent from students' responses in tests, for example, that they often did not make the required sense of word problems. Cynthia's response to the Mrs Patel problem was not atypical. The teacher therefore developed various activities with the aim of supporting students to make sense of word problems, including the task of writing word problems of their own. I became interested in this particular task because I noticed that it generated a good deal of engagement and interaction on the part of students, including the bilingual students. I wanted to know what was going on in these engaged discussions, so I recorded the bilingual students as they worked on the task. In all, over a period of two years, I collected 28 such recordings involving 10 different bilingual learners, all of which I transcribed. All the participants were able to write several problems together, though not all students were able to solve all of their problems.

One of the challenges of investigating how bilingual learners make sense of mathematics relates to the issue of working across languages that I have already mentioned. How can I investigate what sense Cynthia, for example, makes of word problems? I could ask her, but this seems problematic. We would use the same words, but the meanings would be different. Our experiences and understandings of milkshakes and sandwiches and Mrs Patels will be quite different, as will our experiences of word problems, mathematics and school learning. In analysing the recordings, therefore, I needed an approach that did not make assumptions about what the participants were 'really' thinking. I developed an approach based on discursive psychology (Edwards & Potter, 1992; Edwards, 1997). This approach entailed detailed analysis of the students' interaction. Crucially, however, I was interested in *how* the students talked, rather than in what they meant. By examining how the students discussed, agreed, disagreed and negotiated their work together it is possible to say something about what is relevant to the students as they write word problems and so about how they *explicitly* make sense of such problems. My analyses suggest that the task of writing word problems was, as the teacher hoped, successful in supporting her students to develop a deeper understanding of how word problems work. Moreover, the task was particularly supportive of bilingual learners, as I show in the next section.

Writing Word Problems: Engaging Bilingual Learners

What, then, did all these recordings reveal? In general, they showed that these bilingual learners were able to participate fully in the task of writing word problems. The recordings are notable for the amount of discussion that took place. Some pairs or groups worked in a fairly individualistic way, usually taking turns to produce a word problem. More typically, however, the students entered into sometimes lengthy discussions, debating and negotiating various aspects of their problems, as well as producing solutions. Furthermore, the students' discussions showed that a good deal of attention was paid to several important aspects of word problems, including their form, their mathematical structure and the language used (Barwell, 2005a). It was evident, therefore, that these bilingual learners were able to make sense of word problems and that this sense making supported their mathematical thinking. In this section, I summarise three ways in which the word problem task appeared to promote students' sense making. First, the task allowed students to bring personal experiences into their thinking. Second, the task led students to pay attention to the nature of the word problem genre. Third, the task allowed for a productive, mutually supportive interaction between language learning and mathematical thinking. These three aspects of the students' sense making contributed to the development of meaningful mathematical thinking in different ways. I will illustrate and discuss these different kinds of sense making, drawing on transcripts of Cynthia's discussions with a monolingual student called Helena (for a full analysis of the transcript, see Barwell, 2003a). I have chosen to refer to one recording of these two students to maintain a degree of continuity in my presentation of these ideas. Many other examples involving some of the other bilingual learners could also be used to illustrate some or all of the same points. Indeed, many of the ideas presented here come from analyses of other transcripts, which I indicate with references to the relevant work. (See Note 1 for transcript conventions.)

Drawing on personal experience

If students generally struggle to connect the real life context used in a word problem to the mathematical task, the students who participated in my research appeared to have little difficulty making such connections. A key difference, however, is that, in writing their word problems, the students made reference to real life experiences of their own. These references to personal experience were frequently developed into contexts for the students' word problems, or used to substantiate ideas for elements of their problems. In the following series of extracts, Cynthia and Helena prepare and solve one word problem. Their discussion begins with Helena asking Cynthia what she likes:

Extract 1
345   H: Cynthia what d'you like
346   C: I like/ my mum
347   H: okay then/ Cynthia has fifty pounds/ to buy her mum a present

348 C: (*laughs*)
349 H: and she gets her/ a big dress/
350 C: big dress/ no/ my mum doesn't like dress/ I get her ahhh/ big music box/
351   if you open it/ [it's music/
352 H: [a stereo
353 C: no/ if you open it/ it's music/ it's music and you have put jewellery in/
354   [like/ like/ yeah
355 H: [oh/ oh/ that that/ you open it and it goes beep ooo/
356 C: music

Helena takes Cynthia's statement that she likes her mum and proposes the start of a word problem about Cynthia buying her mum a present. Cynthia's personal experience is used in several ways in this extract. Most obviously, writing a problem about Cynthia and her mum introduces a personal connection. What is perhaps more interesting, however, is the way personal experience is used to *reason* about the problem they are writing. Thus, Cynthia rejects the suggestion that Cynthia in the problem buys her mum a dress by saying 'my mum doesn't like dress'. Similarly, in the next extract, Cynthia argues about the cost of a 'music jewellery box' by saying 'no I saw one':

Extract 2
371 C: yeah three voh/ and I bought her/ um/ I bought her twenty five pound a
372   music/ jewellery box
373 H: no it only cost/ sixteen pounds/
374 C: [no I saw one
375 H: [and you buy her something
376 C: no/ I I saw one/ it's thir-thirty pounds/ three voh/ no one three/ one three
377   pounds I saw one
378 H: thirteen
379 C: one three/ on that shop/ you have to look at the magazine/ that they that
380   the shop magazine/ and you choose a number and you write it down/ and
381   you gave them/ and and that one you pay the money and you have to
382   wait and they gave you/ you know that shop/ I saw it's just thirty pound/
383   three/ thir-**teen** pounds/ one three/

A second use of personal experience apparent in these two extracts is to clarify or negotiate linguistic meaning. In Extract 1, for example, the two students exchange accounts of their experiences of looking at music jewellery boxes in order to negotiate what is being referred to. Helena, for example, mentions the sound it makes when you open it, 'beep ooo'. More elaborately, in Extract 2, Cynthia backs up her claims about the price of a music jewellery box by giving an extended description of visiting a shop and looking up the object in a catalogue. This kind of description is a good rhetorical strategy; it is more difficult to argue with claims based on this kind of specific experience (Edwards & Potter, 1992).

The way in which Cynthia and Helena use references to their personal experience is typical of the discussions I recorded, although it occurred in some recordings more than others. Sometimes students threw in a brief reference to personal experience to back up a suggestion. In one recording, for example, Safia supports the idea of a problem about 1000 people being at the furniture store Ikea by saying 'last time I went there it was too crowded'. By contrast, some recordings feature extended discussions of personal experiences. In one recording, two students engage in a lengthy comparison of the spending and shopping habits of their mothers. In another, in which Cynthia is working with two other students, she gets into a long discussion about how much pocket money they receive. The discussion develops a slightly edgy feel when Cynthia talks about not getting any pocket money if her father is unemployed, in contrast to the more affluent accounts given by her partner (see Barwell, 2005a). While these accounts sometimes seem like 'off-task' talk, they are often implicated in the students' sense making in relation to their word problems. In the pocket money discussion, for example, the two students are writing a word problem about pocket money. While their discussion and comparison of experiences does not feature explicitly in their word problem, it does provide it with a rich contextual foundation. The words 'pocket money' in their problem are not almost empty signifiers, as may often be the case in textbook problems. Rather, through their discussion, the words develop complex, nuanced meanings based on the shared exchange of experiences. Similarly, Cynthia and Helena's discussion contributes to the construction of a more meaningful word problem.

Attention to the word problem genre

Since students were asked to write word problems, it is perhaps not surprising that they paid attention to their generic features; such attention would ensure that they produced problems of the appropriate form. This attention is already apparent in the previous two extracts. Indeed, Helena's first suggestion is the generically highly appropriate 'Cynthia has fifty pounds/to buy her mum a present'. This suggestion provides a good minimally described scenario, as we would expect at the start of a word problem. Genre surfaces more explicitly in the following extract:

Extract 3
408 H: Cynthia has thirty pounds for/
409 C: no/ not for her her mum/ if (I bought)/ for my mum
410 H: for her mum's present
411 C: if give my mum thirty pound I bought nothing from her/ that not make
412   sense

```
413  H:  no/ I won't writing for you mother/ I said Cynthia has thirty pounds for
414      her mother's present
415  C:  thirty pound/ I gave thirty pound for my mum present
416  H:  no/ I didn't say give it to her
417  C:  then how why you
418  H:  you have thirty pounds [ for your mum's present
419  C:                        [ no
420      but/ I think this make sense/ Cynthia has thirty/ pound/ thirty pound/ she
421      bought err something something something/ it's cost something
422      something/ from her mum present/ and how much she left?/ is that make
423      sense little bit
```

In this exchange, the two students are negotiating wording and linguistic meaning, with, it appears, a feeling that they are not quite understanding each other. This situation leads up to Cynthia's attempt to clarify what she means (lines 419–423), in which she enunciates what amounts to a template problem. Her statement contains all the key elements of a word problem described by Gerofsky (1996): a set-up, some information and a question. It also has 'blanks' where the details can be added, conforming with Gerofsky's observation that these kinds of details are interchangeable. Cynthia's assertion of how she sees the problem they are writing gives us her explicit interpretation of what a word problem looks like. Its generic suitability suggests that she has considerable experience of word problems.

The linguistic features of their word problems were something that all participants paid attention to during their discussions. Most discussions began with consideration of a set-up. Often these set-ups were based on a place, such as a furniture store, a concert, a swimming pool or a morgue. In other cases, they involved named people (real or invented). Participants also included suitable information, including numerical details, as well as concluding questions or tasks. In general, the inclusion of these various kinds of information was not questioned, suggesting that they were expected elements in word problems. Of course, not all the students' word problems were ideal: one pair, for example, included the question 'how many can you divide' and some did not contain enough numerical information for a calculation to be possible. Nevertheless, all the problems included the three components identified by Gerofsky (1996) in some form.

As with the students' accounts of personal experience, their attention to the features of their word problems contributes to the sense-making process. The often implicit three-part structure for the problems provides a degree of structure for the students' work. Indeed, the form of word problems appears to be the primary structure the students draw on: at no point did any of the students begin by discussing a mathematical structure, for example, such as a division relationship. Their discussions began with negotiation of the set-up. There was often a good deal of discussion about the particularities of each problem, such as which location, how many people or how much pocket money, but these discussions took place within the taken-for-granted structure of the problems.

Occasionally, explicit discussion of the form of their word problem does occur, as in Extract 3. It is interesting that Cynthia's assertion of how she understands the problem should come at a point where the two participants are working hard to understand each other. Explicit attention to the structure of the problem, therefore, is implicated in the negotiation of meaning, so contributing to the construction of a meaningful problem more generally.

Interaction between language learning and mathematical thinking

All the bilingual participants in the recordings were in the relatively early stages of learning English. There were times when this became apparent, such as in Extract 3, where Cynthia and Helena appear to talk at cross-purposes. Their debate particularly concerns the use of the word 'for'. Cynthia makes clear her interpretation that having 'thirty pounds for her mum's present' means that she gives the money directly to her mum. Both participants try various reformulations until Cynthia proposes her own version of the problem, a version that avoids the use of 'for' in the opening sentence. This exchange is an example of the regular negotiation of what particular words mean and can be seen as contributing to Cynthia's learning of English. Other aspects of English were also negotiated, including pronunciation, spelling, punctuation and tense. In the next extract from the transcript of Cynthia and Helena, for example, the pronunciation of fifty and fifteen becomes an issue:

Extract 4
444 H: what else would you buy
445 C: wait wait wait wait wait/ bought music box/ it's cost/ it cost/
446 H: how much
447 C: any you like fifteen or thirteen thirteen and
448 H: it costs
449 C: fifteen
450 H: fifteen pounds
451 C: fifteen pound/ co-/ fifteen pound
452 H: (...)
453 C: fifteen pound and she bought/ she bought err
454 H: and she bought/ Cynthia has thirty pounds she buys a music jewellery
455   box [and
456 C: [and she bought a jew- and she bought a ring from her mum// wait
457   wait wait wait/ you write fifty pound/ um/ um um/ (miss)
458 H: (miss)
459 C: jewellery music box/ is costs fifty pounds/ and she bought/
460   [and she um (...) fifty pound/ from one

```
461  H:  [and
462  C:  and she bought/ and she bought a ring/ because I've got/ tw- I've got
463      fifty pound left
464  H:  she bought a what
465  C:  a ring
466  H:  a ring?/ for fifty p- fif-teen pound
467  C:  no/ you know?/ I've got forty p- thirteen pound innit?/ and bought this/
468      and I got fifty pound left/ [ and I bought
469  H:                              [ not fifty
470  C:  one five how d'you say fif-teen/ no fifteen/ fifteen pound/ she's got fifty
471      I've got fifty pound there/ and I bought a ring/ from my mum/ and how
472      much/ cost everything
```

In this extract, Cynthia is heard by Helena as saying 'fifty', something that she finds problematic. The situation is resolved, first by Helena highlighting the alternative possible pronunciation 'fifteen', and then by Cynthia reading out the digits of the number she means. Again, this negotiation can be seen as contributing to Cynthia's learning of English (see, e.g. Swain & Lapkin, 1995).

There were many other examples of discussion that could be related to language learning during the recordings. In my analysis of a recording of two students from Pakistani backgrounds, for example, I noted the following practices in a short sequence in which they debated whether to use the word 'give' or 'gave' in their problem:

- asserting one of the two words;
- offering a version of the problem containing one of the words;
- writing a version of the problem;
- spelling aloud;
- reading the problem back;
- revising the written version of the problem. (Barwell, 2005c: 215)

I do not wish to simply show, however, that the word problem task provides opportunities for language learning. These language-focused exchanges are not separate from the students' mathematical thinking. The significance of these exchanges is that they contribute to the development of the mathematical dimension of the students' word problems. In Cynthia and Helena's problem, for example, the difference between fifty and fifteen is crucial to the mathematical structure of the final problem. Similarly, the discussion in Extract 3 relating to the word 'for' concerns a central aspect of the mathematical structure of the problem, since using money to buy a present is a 'real life' representation of a mathematical relationship between the amount of money and the value of the present. By negotiating what their words mean, Cynthia and Helena are also negotiating the mathematics of their problem. Conversely, the emerging mathematical structure for the word problem provides the basis for the meaning of the words that the students are using and sometimes explicitly discussing. Hence, mathematical thinking and language learning cannot be seen as two separate processes; they are inseparably interwoven (Barwell, 2005c).

Supporting Mathematical Thinking

I have summarised three aspects of bilingual students' sense making about word problems that have emerged from my analyses of their participation in the word problem task. The students used accounts of personal experience, attention to generic form and negotiation of linguistic form and meaning as they worked together to write their word problems. All three aspects are implicated in the students' mathematical thinking in relation to their word problems. The link between them is not necessarily a simple one; it is not generally possible for me to identify a specific bit of personal experience that goes with a particular instance of mathematical thinking. Rather, these ways of making sense of word problems contribute to the development of richly contextualised, meaningful problems. The problems written by the students were generally as concise as word problems usually are. It may be that, for many students, particularly students learning the language in which the problems are written, the conciseness of word problems makes them difficult to get hold of. For the students who took part in my research, however, the conciseness of their word problems marks the *end* of a process of sense making, rather the beginning of such a process. Many of the different elements of the problem have been discussed and sometimes revised, building up a set of interconnected meanings. In many cases, solving the problems was then relatively straightforward.

In the case of Cynthia and Helena's word problem, the final problem looks like this:

> Cynthia had £30, she buys a music jewellry box,
> costing £15 and she buys a ring costing £12.99
> How much does she spend altogether and How
> much does she have left?

Their lengthy discussions have resulted in a highly suitable example of a word problem. Interestingly, there are no references to Cynthia's mum or the purpose of buying the items mentioned. Indeed, these omissions make the problem more suitable: word problems do not usually include motives. This is not to say, however, that these elements are absent from the students' interpretation of the problem, since, as the transcript shows, they have discussed them carefully. For Cynthia and Helena, buying presents for Cynthia's mum is part of the context of the problem. In a similar way, the word 'for' does not appear in the problem, but the discussion that the two students had around this word also contributes to the context of the problem. Their discussion helped to clarify what was going on in the problem. Thus, Cynthia and Helena's discussion creates a rich context for their problem that is not apparent in its final wording. This context is closely linked to the mathematics of the problem. Clarifying what 'for' means, for example, is also about clarifying that the problem is a subtraction. Sorting out the difference between 'fifty' and 'fifteen' concerns the mathematical viability of the problem. Within this meaningful mathematical context, Cynthia, using a calculator, solves the problem with little difficulty.

Extract 5

```
506  C:  yeah how much (...) left/ okay/ do it now/ come on/ no no no/ do that/
507      um/ fifteen and/ one two nine nine and one five oh oh/ okay/ one/ no
508  H:  just like fifteen and twelve
509  C:  no/ I've got you've got twelve pound ninety nine/ twelve nine nine/ take
510      away/ one five oh oh/ eq-/ no/ not [ take away/ it's add/
511  H:                                    [ no not take away/ add
512  C:  two oh nine nine/ add/ one five oh oh/ two seven nine nine/ two seven
513      nine nine/ and three oh oh oh/ take away/ two/ seven nine nine/ equal/
514      two pound and one p./ how much she spent
515  H:  she spent
516  C:  yeah/ wait wait
517  H:  twenty seven ninety nine
518  C:  (...) spent/ S PE N/ she spent/ twenty seven pounds and ninety nine p./
519      left/ and/ she left/ shu left/ she left/ um/ two pound and one p./ done
520      it/ mister Barwell
```

At first, Cynthia starts to perform a subtraction, perhaps echoing her response to the Mrs Patel problem. In this case, however, she is quickly aware that subtraction does not make sense and changes to using addition. In the last part of Extract 5, Cynthia clearly relates the arithmetic calculations to the context of the word problem.

Teaching, Word Problems and Bilingual Learners

Writing word problems allowed the participating students to do a number of useful things. First, it allowed the students to deepen their familiarity with how word problems work. If students understand how the problems work, they are likely to be in a better position to make sense of problems they have not seen before. Second, the task provided space for students to bring their own lives and worlds into the task. It is worth noting here that the teacher did not need to know about the students' backgrounds. In such diverse classrooms, it is impractical for teachers to be expected to learn about all their students' backgrounds in any depth. The word problem task, however, provided a space in which the *students* could draw on their own experiences and incorporate them into their work. Third, the task promoted language learning in a mathematical context. The value of this aspect of the task is that language learning is fully integrated with mathematics learning. Finally, the task led to students making links between the three preceding points. One of the challenges of word problems is to be able to link the language, the scenario and the mathematics. In writing word problems of their own, this is what Cynthia and her fellow students were doing.

The four ways in which the word problem task was supportive of bilingual learners are not exclusive to any particular task and can be used to guide the development of other tasks for bilingual learners. Other ideas for working with word problems might include: comparing word problems; modifying word problems and discussing what difference the modifications make; taking a previously unseen word problem and writing a story about it or drawing a picture to illustrate it; taking a previously unseen word problem and rewriting so that it involves familiar people or places but retains the same mathematical structure (see Barwell, 2003b).

In England, bilingual learners will be expected to solve word problems in tests for the foreseeable future. Students like Cynthia will, therefore, need to be able to look at problems about Mrs Patel and have a sense of what they need to do. As Mestre (1986) implies, teaching Cynthia word problem vocabulary is not likely to be sufficient support for her to be able to respond successfully. Cynthia's teacher provided her with a task that enabled Cynthia to engage with word problems, to engage with their form, their language and their mathematical structure, and thereby to learn both mathematics and English.

Acknowledgement

I am grateful to the teacher and students who took part in the research referred to in this chapter.

Note

1. Transcription conventions: Bold indicates emphasis. / is a pause <2 secs. // is a pause >2 secs. (...) indicates untranscribable. ? is for question intonation. ( ) are for where transcription is uncertain. [shows concurrent speech. Capitalised italics indicate sounded out letters during spelling: *T* denotes 'tee' for example. All names are pseudonyms.

Chapter 6
How Language and Graphs Support Conversation in a Bilingual Mathematics Classroom

JUDIT MOSCHKOVICH

In this chapter, I examine a mathematical discussion about two graphs that took place in an eighth grade (13–14 years old) bilingual mathematics classroom. The discussion involved multiple ways of seeing and talking about the scales on the vertical axes. In particular, the teacher and students used multiple meanings for the phrase 'I went by ...' to describe the scales (e.g. 'I went by ones' or 'I went by twos'). The graphs, verbal descriptions, gestures and the multiple meanings generated during the discussion were all resources for socially constructing interpretations of the graphs. I use a situated perspective on learning and discussing mathematics. I use the term 'situated' to mean 'local, grounded in actual practices and experiences' (Gee, 1999: 40). More specifically, I take a situated view of language and meanings: I assume that in all discussions, whether they occur in one or two languages or among native speakers of a language, the meanings for utterances are situated or grounded in the local situation. I also assume that participants interpret utterances and representations in multiple ways. I view these multiple interpretations as part and parcel of sense making in the local situation, rather than mistakes. Lastly, rather than focusing on obstacles, I focus on describing how utterances and representations provide resources for discussions.

Using this perspective, I address the following questions: What are students' multiple interpretations of graphs? What kinds of resources do bilingual students use to discuss graphs? I use transcript excerpts to illustrate multiple meanings for the phrase 'I went by' and multiple views of the scales on two graphs.

Multiple Interpretations in Bilingual Mathematics Classrooms

Why focus on mathematical discussions that involve multiple interpretations, meanings and ways of talking? Researchers have suggested that an important function of productive classroom discussions is to bring different ways of talking and points of view into contact (Ballenger, 1997; Warren & Rosebery, 1996). Such discussions may be particularly important in mathematics instruction because they contradict pervasive beliefs about mathematical activity, such as the idea that there is no room for interpretation (Borasi, 1990).

A focus on mathematical discussions is particularly important for bilingual classrooms. On the one hand, we could imagine that mathematical discussions would be difficult to create and maintain in bilingual classrooms. After all, are bilingual students not struggling with language? Should we not be concerned that an instructional emphasis on mathematical discussions will make bilingual students look less competent than traditional computational work? The discussion about graphs examined below is a counter example to the imagined difficulties that bilingual students might face in discussing mathematical concepts. This mathematical discussion shows that bilingual students are, in fact, able to participate in mathematical discussions that are conceptual and involve multiple interpretations. The question is not *whether* bilingual students can engage in these types of discussions but *how* instruction can support bilingual students in participating in discussions and in learning to communicate about mathematical concepts. Too often, descriptions of bilingual students focus on the obstacles they face in understanding text or utterances in English and these misunderstandings are invariably ascribed to their lack of proficiency in their second language. In contrast, the discussion shown below shows that multiple interpretations, rather than being seen as caused by language difficulties, can be seen as reflecting how a student was grappling with a mathematical concept.

Linguistic and Educational Setting

Carlos and David are students in an eighth-grade bilingual class in an urban area in Massachusetts in the United States. They are both native Spanish speakers who are bilingual. In their school, there is a 'two-way' or 'dual immersion' bilingual programme for Grades K–6. This means that in Grades K–6 (5–12 years old), students spend half their instructional time in English and half in Spanish. In Grades 7 and 8 (12–14 years old), classes are no longer two-way bilingual. Instead, teachers and students use both languages depending on the setting and participants. Most of the students in Carlos and David's class have been in the programme for several years, many since elementary school. Some of the students are recent immigrants, several students are Spanish dominant and most students are proficient in both Spanish and English.

Carlos and David arrived in the United States from Central America as young children and both have been in the bilingual programme since the early grades in elementary school. They report sometimes speaking Spanish at home, and in the classroom they seem to switch easily and fluidly between monolingual and bilingual modes (Grosjean, 1999). When discussing a mathematics problem together, they will throw in words, phrases or extended talk in Spanish. When talking to the teacher, they tend to use mostly English. Thus, they represent an important and significant segment of the Latino/a population in the United States: those students who would *not* be labelled as Spanish dominant.

The class was conducted mostly in English, with some discussions and explanations in Spanish. The teacher used Spanish mostly when addressing students who were seen as Spanish dominant. Some students spoke mainly English, some students used both languages and some students spoke mainly in Spanish.

This is a classroom where students expect to make sense of their work, discuss their work with peers and also use the teacher as a resource in their discussions. Students took on some of the responsibility for explaining and understanding solutions. Students engaged in serious and extended discussion of their solutions. The group discussions seemed to be important to the students. Nevertheless, while the students shared responsibility for explaining solutions, they sometimes also tended to rely on the teacher as the authority for evaluating a solution.

In the next section, I examine a discussion between Carlos and David. Although both students are bilingual, I have selected a discussion that transpires in only one language, English, on purpose and for several reasons. First, their discussion reminds us that many conversations in bilingual classrooms take place in only one national language. Second, and perhaps more importantly, their discussion highlights how multiple interpretations are not tied to the use of more than one national language but are connected to the negotiation of mathematical meanings. And lastly, I chose this example because, in the United States, bilingual Latino/a students who are labelled as 'English dominant' represent an important and significant segment of the Latino/a population. I will use the discussion between Carlos and David to ground deeper consideration of these broad issues. First, however, I need to present their discussion in some detail in order to make visible what is taking place, what each participant means and where the discussion is going (a more detailed analysis of some aspects of this transcript appears in Moschkovich, 2008).

The transcript comes from a larger set of data collected in Carlos and David's class. Classroom observations and videotaping were conducted during two curriculum units from *Connected Mathematics* (Lappan *et al.*, 1998), 'Variables and Patterns' and 'Moving Straight Ahead'. The discussion involving Carlos and David occurred during the unit 'Moving Straight Ahead'. Data collected included video recordings of whole-class discussions and at one student group for every lesson, as well as videotaped problem-solving sessions in pairs.

Carlos and David's discussion occurred towards the beginning of a classroom period. The teacher usually started the 90-min class with a brief whole-class discussion about a mathematics problem. Students then worked in groups of two to four, discussing the problem at their tables. The teacher moved from small group to small group, asking and answering questions in each group. Towards the end of the class period there were usually reports or presentations by each group as well as whole-class discussions led by the teacher. On the day of this discussion, students expected that each group would at some point be asked to go to the front of the classroom to explain their graphs or charts and would be expected to describe how and why they solved a problem as they had, as well as be prepared to answer questions from other students and the teacher.

Making Sense of Two Graphs

The class had been working on several problems about a five-day bicycle tour. In the story, while some riders rode bicycles, other riders rode in a van and recorded the total distance from the starting point for the van every half-hour. The problem below (Figure 6.1) refers to the second day of the tour.

Carlos and David often worked together in a small group. We join their discussion as they compare their answers to this problem and review the graphs each created independently for homework. Carlos first read his written answer. Then David read his answer. Carlos and David agreed in two ways in their written answers. First, they both wrote that the bikers travelled a total of 45 miles in 5 h. Second, they both wrote that the half-hour time interval in which the bikers made the least progress was the one between 2.0 and 2.5 h. However, they did not write the same answer for which half of the trip they made the most progress and for the half-hour interval during which they made the most progress.

As the discussion began, David and Carlos noticed that they had different answers for this problem. They looked at their graphs and noticed that these graphs looked different (Figures 6.2 and 6.3). When reading the transcript, it is important to continue to look at the graphs to understand

On the second day of their bicycle trip, the group left Atlantic City and rode five hours South to Cape May, New Jersey. This time, Sidney and Sarah rode in the van. From Cape May, they took a ferry across the Delaware Bay to Lewes, Delaware. Sarah recorded the following data about the distance traveled until they reached the ferry.

Time (hours)	Distance (miles)
0.0	0
0.5	8
1.0	15
1.5	19
2.0	25
2.5	27
3.0	34
3.5	40
4.0	40
4.5	40
5.0	45

1. Make a coordinate graph of the (time, distance) data given in the table 2. Sidney wants to write a report describing the day 2 of the tour. Using information from the table and the graph, what would she write about the days travel? Be sure to consider the following questions:
   A. How far did the group travel in the day? How much time did it take them?
   B. During which interval(s) did the riders make the most progress? The least progress?
   C. Did the riders go further in the first half or the second half of the days' ride?
2. By analyzing the table, how can you find the time intervals when the riders made the most progress? The least progress? How can you find these intervals by analyzing the graph?

Figure 6.1 Problem: From Atlantic City to Lewes (*Connected Mathematics*, Lappan *et al.*, 1998)

what Carlos and David are referring to and to get a sense of the discussion. What aspects of the graph is each participant focusing on? (For transcription conventions, see Note 1.)

Episode 1: David and Carlos compare their graphs

22	**David:**	Here's my graph. Did it come out like yours?
23	**David's gesture:**	((Turning his paper towards Carlos.))
24	**Carlos:**	I don't know.

25	Carlos's gesture:	((He looks at both papers and compares graphs.)) ((0.2 sec.))
26	David:	I'll do mine. =
27	Carlos:	= It's because you did it upwards.
28	Carlos's gesture:	((Sweeping his right hand, with his finger pointing up.))
29	David:	Oh, you did. ((0.2 sec.)) ((Teacher approaches the group.))
30	Carlos:	Were we supposed to do the graph upwards? Or to-.
31	Carlos's gesture:	((Sweeping his hand upwards again and then moving his hand horizontally in the air.)) [
32	David:	Or, or or crooked like this? Whatever.
33	Carlos:	Or horizontally? ((0.2 sec.))
34	Teacher:	Doesn't matter. ((shaking her head))
35	Carlos:	'Cause like, like when we look at our graphs his is going up and mine is going towards the side.
36	Carlos's gesture:	((Moving his right hand upwards and then moving his right hand horizontally in the air.)) ((0.2 sec))
37	Teacher:	Did you have the same, did you put both of the same things on the x-axis and on the y-axis?
38	Teacher's gesture:	((Looking through students' graphs.))
39	Carlos:	No, ((0.2 sec)) yes, actually.
40	Teacher:	You both put time here. OK, so what's different about these? I don't think it's, I don't think it's the positioning of them. Look at your numbers, the way you placed your numbers.
41	Teacher's gesture:	((Points at the x-axis on both papers, back and forth. Then turns David's paper in the same direction as Carlos's paper.))

In this first episode, Carlos and David began the discussion of their solutions to the homework problem by considering whether their graphs were the same or not. They apparently seemed to share the expectation that their graphs would or should look the same. The teacher then joined their group. Carlos asked her how they were supposed to do the graph. She responded first by asking them to compare their axes (Line 37). The teacher's participation in the discussion began by addressing the issue of what is the same and what is different about the two graphs. She first considered whether David and Carlos had the same variable assigned to each axis (Line 37) and concluded that they both had put time on the

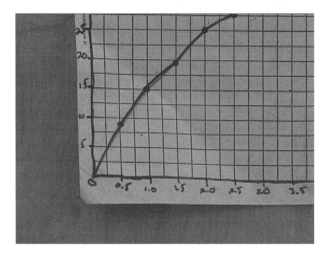

Figure 6.2 Carlos's graph

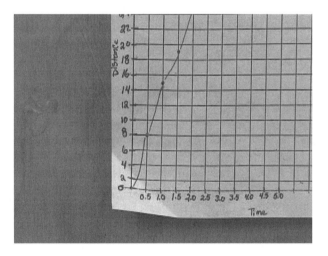

Figure 6.3 David's graph

x-axis. Next she considered what was different about the two graphs and suggested that it was the way that David and Carlos had 'placed their numbers' (Line 40).

In Episode 1, the teacher set the goal of finding similarities and differences between the two graphs. The students seemed to be focusing on the graph at a global level, describing the overall shape of the curve. In contrast, the teacher defined the goal as sorting out 'sameness' and

How Language and Graphs Support Conversation

'difference' at a local level focusing on the axes, not the shape of the curves. She did this through the following questions and statements: 'Did you put both of the same on the x-axis and the y-axis?', 'You both put time here (on the x-axis)', 'So what's different about these?' She thus shifted the discussion and the students' focus from the 'positioning' of the graphs (whether each graph goes 'upwards' or not, as described by the students in Lines 27 and 30) to the way each student had placed numbers on the axes.

In the next episode, the teacher considered the impact of the scale on the shape of the graph, using a global view of the graphs, and then began to construct and share her descriptions of the students' scales on the y-axis. Carlos began to describe how he labelled the axes of his graph (see Figure 6.1) making tick marks every two segments. In contrast, David had labelled the axes of his graph (see Figure 6.2) with tick marks on every one segment. Again, when reading the transcript, it is important to continue to look at the graphs to understand the referents of utterances. In the next episode, what does each participant mean by the phrase 'I went by'?

Episode 2: Describing the shape of the curve

42	**Carlos:**	Oh, that's true, 'cause I went by twos, I went 1, 2 ((0.2 sec)) and then I put that one (---) he went by one.
43	**Carlos's gesture:**	((Begins counting with his right finger following the numbers on his paper.))
		[
44	**Teacher:**	Aha, You skipped one (referring to a segment on the scale on Carlos's graph). So how does that change how it looks?
45	**Carlos:**	'Cause it doesn't go up as far, it only goes, it's more steeper. It looks more steeper.
46	**Carlos's gesture:**	((Moving his right hand outward. Then moving his right hand straight up.))
47	**Teacher:**	Remem-. Similar to the difference between this one and ((0.1 sec.)) and this one here. Right?
48	**Teacher's gesture:**	((Makes a sign with right thumb and index of her hand to show interval differences on their papers. Then she points to a graph on the blackboard. Next she points to a second graph on blackboard. The first and second graphs have different scales on the x-axis so that the second graph is compressed along the x direction.))
		[
49	**David:**	That one.

50	**Carlos:**	Yeah.
51	**Teacher:**	Here the numbers are closer together so it looks looks steeper. Other than that are they the same graph?
52	**Teacher's gesture:**	((Makes a sign with thumb and index finger. Then she gestures upwards with her right hand.)) ((0.1 sec.))
53	**Carlos:**	No, also here in the x-axis.
54	**Carlos's gesture:**	((Carlos point to the x-axis on his paper.)) [
55	**David:**	(the distance) ((Points to the axis on his paper.))
56	**David:**	I went by twos.=
57	**Carlos:**	= This is the x-axis. Right?
58	((Carlos points to the axis.))	
59	**Teacher:**	This is the y-axis, ((0.1 sec.)) this is the x-axis.
60	**Teacher's gesture:**	((Sweeps her pencil vertically to represent the y-axis and then horizontally to represent the x-axis))

During Episode 2, Carlos introduced the phrase 'I went by twos' (Line 42) to describe his own y-axis scale and the phrase 'he went by one' (Line 42) to describe David's scale. Carlos continued to use these phrases during Episodes 2 and 3 as he described how he had labelled the axes of his own graph. Turning to the graphs, we can see that Carlos had labelled his axes by making a tick mark at every two-grid segment. In contrast, David had labelled the axes of his graph making tick marks on every grid segment.

In Episode 3, below, the teacher and Carlos clarify the meanings for 'I went by ...' What are the claims each participant is making about the scales?

Episode 3: Using and clarifying 'I went by ...'

61	**David:**	I went by twos.
62	**Teacher:**	You went by twos and you went b:y- ((0.2 sec.))
63	**Carlos:**	I went by twos. You (didn't) you went by ones! What are you talking about. [
64	**Teacher:**	No, here on the y-axis.
65	**Teacher's gesture:**	((Points to the axis.))
66	**Carlos:**	Oh, I went by fives.
67	**Teacher:**	You went by fives. ((0.2 sec)) No, actually you didn't go by fives. You actually went by two

		and a halves because you'd, you did every 2 spaces was five
68	Teacher's gesture:	((She points to Carlos's paper while she explains.)) ((0.3 sec.))
69	Carlos:	Then he only went by one.
70	Carlos's gesture:	((Carlos points to David's paper.)) ((0.2 sec))
71	Teacher:	Every one space was two of his. You'd see, they're almost the same. If you look at the next two-((puts down her notebook and points to the graphs))
72	Carlos:	[Wait! but I don't get what you're saying.
73	Teacher:	OK.
74	Carlos:	'Cause I went by fives. ((David stands up))

During Episode 3, although David and Carlos had labelled their axes differently, David initially claimed that he also 'went by twos' (Line 61). The teacher first accepted this claim and proceeded to describe Carlos's scale. Carlos disagreed with David, insisting that while he 'went by twos', David 'went by ones' (Line 63). At this point Carlos seemed to notice that they might not all be talking about the same thing, saying, 'What are you talking about?' (Line 63). Carlos then changed the description of his own scale, saying, 'Oh, I went by fives' (Line 66). The teacher first agreed with this claim saying, 'You went by fives' (Line 67), but then, after a short pause, she disagreed, proposing that Carlos had not gone by fives but, rather, had gone by 'two and a halves' (Line 67). In response, Carlos proposed that, if that were the case, then David 'went by one' (Line 69). The teacher explained that on David's scale, each space had a value of two units, 'Every one space was two of his' (Line 71). At this point, Carlos said that he did not understand the teacher's explanation and returned to claiming that he 'went by fives' (Line 74).

With the series of actions and utterances at the end of Episode 3, the teacher showed her commitment to focus on a detailed comparison of the two scales. By putting her own notebook down, looking at the two graphs, pointing to the axes on each graph, touching the papers and orienting the two graphs so that they are facing her, she called on the students to focus their attention on the two y-axis scales. In response to these actions, David reoriented himself, standing up (Line 74), perhaps so that he could read both graphs.

During Episode 4, the teacher responds to Carlos's claim that he 'went by fives.' What is the teacher's role in the discussion?

Episode 4: Teacher responds to Carlos's claim that he 'went by fives'

75	**Teacher:**	OK, your numbers, right, the numbers you have are by five ((0.2 sec)) OK ((0.1 sec.)) If you look at one line here, what number is he at?
76	**Teacher's gesture:**	((Takes David's paper and places it next to Carlos's paper. Then points to David's graph.))
77	**Carlos:**	Two.
78	**Teacher:**	What number would you be at if you had a number here?
79	**Teacher's gesture:**	((Points to Carlos's graph.))
80	**Carlos:**	Three.
81	**Teacher:**	Almost, two and a half.
82	**Carlos:**	Yeah.
83	**Teacher:**	Because that'd be half way to five. OK. ((0.1 sec)) At this point, after 1, 2, 3, he's got 6. For you after three, 1, 2, 3, you'd be at 7 and a half.
84	**Teacher's gesture:**	((She counts the squares with her pencil))
85	**Carlos:**	O:K:. [
86	**Teacher:**	See what I mean? So it's actually two and a half. The numbers you wrote are by fives but since you skipped a line in between, each one is two and a half.
87	**Teacher's gesture:**	((Raises her hand and in the air uses thumb and index to show interval.))

Multiple Ways of Talking

There are at least two perspectives of the students' graphs evident in their discussion. The statement 'I went by twos' can be interpreted as describing the action taken to construct the scale, so that 'I went by twos' means 'I went by two segments' (see Figure 6.4).

It could also be interpreted as describing the quantity represented by the chunk created between two tick marks, as in 'I went by two units'. For David's scale this means 'I made tick marks at every two units' (see Figure 6.5).

In both of these interpretations, there is an actor, 'I', who is constructing the graph. Indeed, Carlos and David repeatedly refer to their scales as 'I went by'. The teacher also initially included an active subject, referring to the scales as 'you went by' (Lines 62 and 67). She then moved to referring to the action as 'you did every 2 spaces was 5' (Line 67). She then moved from referring to the action that constructed the graph to focusing on the quantitative relationship between the spaces on each of the two

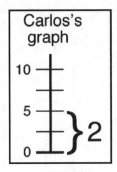

Figure 6.4 'I went by twos' describing number of segments

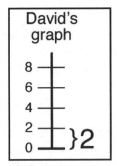

Figure 6.5 'I went by twos' describing number of units

graphs and making a comparison between the two graphs as in 'every one space was two of his (Line 71)'. And lastly, she described the quantitative relationship between the numbers on the scale within one graph, as in 'the numbers you have are by five' (Line 75). Her descriptions were thus moving away from a personal action perspective to an impersonal nominalization based on the numbers and quantitative relationships.

In Lines 75 and 78, the teacher resumed taking a personal action view, saying 'If you look at one line here, what number is he at?' and 'What number would you be at if you had a number here?'. But these were not personal views of the actions taken to construct the graphs. Instead, they were statements about where one would be on the graph if one were moving on the graph at the present moment, some time after constructing the graph.

The teacher alternated between referring to Carlos's actions and to the quantitative meaning of the marks on the scales. She referred to the actions Carlos would take, saying 'you have', 'is he at', 'would you be at', 'he's got', 'you wrote' and 'you skipped' (Lines 75, 78, 83 and 86). She referred to the quantities saying, 'that'd be half way to five', 'it's actually two and a half' and 'each one is two and a half' (Lines 83 and 86). This

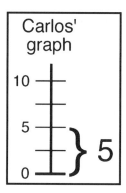

Figure 6.6 Carlos describes his scale as 'I went by fives'

point of view, combined a view of the actions taken to construct the scale with a quantitative view of the meaning of the marks on the scales.

We can see that there were at least three different ways of describing the marks on the axes by looking closely at particular utterances in Episode 4 (Lines 56, 63, 66, and 67). In Line 56, when David said 'I went by twos', he could have been referring either to the value of the interval between labelled tick marks on his graph or to the number of units in that interval (see Figures 6.3 and 6.5). Instead, when Carlos said 'I went by twos' (Line 63) he seemed to be referring to the number of segments between labelled tick marks along the y-axis (see Figures 6.2 and 6.4).

When Carlos said 'I went by fives' (Line 66, see Figure 6.6), he was using a different meaning of the phrase 'I went by', this time referring to the value of the interval between labelled tick marks. Using yet another meaning, when Carlos said that David 'went by one' (Line 69, see Figure 6.7), he seemed to be describing how David had labelled his scale so that tick marks

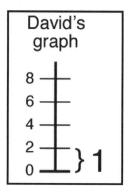

Figure 6.7 Carlos describes David's scale as 'He went by one'

How Language and Graphs Support Conversation

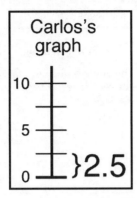

Figure 6.8 Teacher describes Carlos's scale as 'You went by two and a halves'

appear every one segment, thus referring to the number of segments between labelled tick marks. In contrast, when the teacher said 'you actually went by two and a halves', she was referring to the number of units in the interval between tick marks on Carlos's graph (Line 67, see Figure 6.8).

Table 6.1 lists three ways of using 'I (or you) went by' to describe these two graphs. One meaning refers to the value of the interval between tick marks, the second to the number of segments between tick marks and the third to the value of each segment between tick marks. During Episodes 2 and 3, Carlos used the first and second meaning, David used the first and third meaning and the teacher used the third meaning.

Carlos, David and the teacher seemed to be using the phrase 'went by' with different meanings. David seemed to use 'I went by 2' in Line 56 to mean that the value of each tick mark on the y-axis increased by 2. In David's graph, tick marks corresponded to segments so 'two' is also the value of each segment. In Carlos's graph, tick marks did not correspond to segments since Carlos had labelled only every other segment with a tick mark. Not only did the two graphs look different, Carlos also seemed to use 'I went by —' (Lines 42, 63, 66 and 74) in several ways that are sometimes different from David's or the teacher's. On the one hand, Carlos used the phrase in Line 63 to refer to how many segments there were between tick marks on his graph, in this case 2 squares. On the other hand, in Lines 66 and 72, Carlos seemed to be using the phrase to refer to how the value increased for each tick mark on his graph, in this case by 5. There are several ways to interpret Carlos's utterance in Line 69 'Then he only went by ones'. One is that Carlos was referring to how many segments correspond to a tick mark in David's graph. The other is that Carlos was taking the value between tick marks, 2, and dividing by 2 because that was what the teacher had done for Carlos's graph (dividing 5 by 2 to obtain 2.5).

Table 6.1 Multiple meanings for 'went by'

Utterance	Focus of attention	Coordinated utterance and focus of attention
Carlos: 'I went by fives' **David:** 'I went by twos'	The value of the interval between labelled tick marks	'I went by fives' — Carlos's graph (0, 5, 10) } 5; David's graph (0, 2, 4, 6, 8) } 2 — 'I went by twos'
Carlos: 'I went by twos' **Carlos:** 'He ((David)) went by one'	The number of segments between labelled tick marks	'I went by twos' — Carlos's graph (0, 5, 10) } 2; David's graph (0, 2, 4, 6, 8) } 1 — 'He went by one'
David: 'I went by twos' **Teacher:** 'You ((David)) went by twos …' **Teacher:** No, actually you ((Carlos)) didn't go by fives, you actually went by two and a halves, because you did every two spaces as five.	The number of units in the interval between tick marks	'You went by two and a halves' — Carlos's graph (0, 5, 10) } 2.5; David's graph (0, 2, 4, 6, 8) } 2 — 'I went by twos'

During the discussion in Episode 5, the teacher explained to Carlos that each segment in his graph had a value of 2.5. Even after the teacher's explanation, we can see that Carlos used yet another way of describing the graph. Carlos's utterance 'three' in Line 80 is difficult to interpret. If we assume that Carlos knows how to divide 5 by 2 to obtain the correct answer, then his answer that the point on the graph halfway between 0 and 5 is 3, is perhaps evidence that Carlos does not interpret the segments as corresponding to lengths of equal value.

The Teacher's Role in the Discussion

The analysis presented above shows that, during their discussion, the students used phrases of the form 'I went by', which were ambiguous, had multiple and shifting meanings and were coordinated with different views of the scales. In this last section, I show how the teacher did not respond to these multiple interpretations and ambiguous meanings as obstacles. Instead, she used these multiple interpretations as resources: she built on the students' interpretations, highlighted a view of the scale anchored in unitizing and provided students an opportunity to use this view.

In their discussion, the students first took some of the responsibility for understanding their solutions and then invoked the teacher's authority to resolve a question about their graphs. How did the teacher approach this question? Initially she said, 'they (the graphs) are the same'. That did not seem to work for the students. Next, she engaged students in making sense of their graphs. To support the mathematics in this sense-making activity, she used comparisons of equal intervals. The teacher did not evaluate student products as 'right or wrong'. Instead, she engaged the students in a discussion detailing and connecting different interpretations. The responsibility for participating, explaining and understanding was distributed across the students and the teacher. The teacher and the students together participated in a mathematical discussion connecting different interpretations of the scales and multiple meanings of the phrase 'I went by ...'

How did the teacher participate in this discussion? The teacher evaluated how the graphs were the same or different from at least two different perspectives. She then made sense of the two graphs from her own perspective, described to the students how she saw and interpreted the scale and tick marks on the graphs, and then compared the scales on the two graphs. The discussion revolved around what each segment or tick mark represented for each student.

Although the teacher was in a place of authority, she did not produce another graph or evaluate the graphs or the scales. Rather, she based the discussion on the students' work. The teacher did not correct an error, contradict a misconception or provide the one right answer. Instead, she treated both graphs as correct and accepted a situation where there might be multiple interpretations and meanings. Although the teacher contested Carlos's description of his scale 'No, you actually didn't go by fives', she *also* accepted his interpretation saying 'OK, your numbers, right, the numbers you have are by five'. The teacher did not explicitly define what 'went by' meant. Instead, the discussion revolved around what quantity each segment or tick mark represented to each student. The teacher thus supported multiple ways of describing the scales and multiple meanings.

The teacher also used several mathematical concepts as resources. First, she set the goal of looking at the scales rather than looking at the shape of the curves. Initially, the students focused on the curves as objects, comparing the curves in terms of their overall shape, saying that a curve went 'upwards' (Lines 27 and 30) or 'up' (Line 35), in the case of David's graph, or that a curve was 'crooked' (Line 32) or 'towards the side' (Line 35), in the case of Carlos's graph. The teacher proposed and set a new goal, describing the axes rather than describing each curve. This new goal involves implicit knowledge about scales, mainly that the two scales will have an effect on how the slope of a curve appears on the graph. The teacher thus helped the students shift from a view of the curves as objects to a focus on the scales.

The teacher also used the concept of unitizing as a resource. Her descriptions not only moved away from describing a personal action taken to construct the scales towards nominalizations, but they also focused on making comparisons among quantities. In her descriptions, she made a distinction among labels, quantities and measures. The teacher explicitly distinguished between the labels that go 'by fives' and the value of the grid segments as a unit, saying 'You actually went by two and a half' (Line 67). In the second case, the phrase 'you went by' refers to the unit value of one grid segment and is thus an instance of unitizing. The teacher also compared the values of the grid segments on the two scales (Line 71), again an instance of unitizing (Lamon, 1994, 1996).

Lastly, the teacher provided the students an opportunity to use a *unitized* view of the marks on the scales. She set a new problem, determining the value of the *y*-coordinate on each graph after moving up one grid segment on the *y*-axis (Lines 75–82) and after moving up three grid segments on the *y*-axis (Line 83). As she and Carlos jointly estimated the *y*-coordinates on the two graphs, she actively engaged him in talking about and viewing the scales from a unitized point of view.

The students were not passive receivers of an explanation, they were active participants in this discussion. How did the students participate in the discussion? Carlos's active engagement is probably easiest to see. He was involved in responding to the teacher's explanations. First, he re-examined his own graph, saying 'I went by twos'. Then he tried to use the teacher's description of David's scale as 'going by twos' but he was not convinced: 'You (didn't) you went by ones!' At one point he even interrupted the discussion, saying 'Wait! But I don't get what you're saying' (Line 72). The teacher took Carlos's interruption seriously and responded with a more detailed explanation and a new approach to explain the differences between the scales. Carlos also persisted until the teacher accepted that his description 'I went by fives' could make sense ('OK, your numbers, right, the numbers you have are by five').

Although David was quieter than Carlos, he was still engaged as seen, for example, when he stood up to look at the two graphs that the teacher had turned. Even though David contributed less talk than Carlos, he continued to look intently at the two graphs during the discussion in Episode 4, so he seemed to still be participating in the discussion.

Conclusions

In the four episodes, we see two students and a teacher discussing multiple interpretations of a graph and multiple meanings for phrases. During this discussion, the teacher and the two students talked about what quantities and marks on two scales represented. The students and the teacher brought several different ways of talking about the scales on the graphs. The discussion involved different meanings for the tick marks on the axes and different meanings for the phrase 'I went by ....'

The teacher used several instructional strategies to support student participation in a mathematical discussion: she used student-generated products, she used gestures and objects to clarify meaning, she accepted and built on students' responses and she used a central mathematical concept, unitizing. Although the teacher described how she saw the two graphs and the scales, she did not explicitly address the multiple meanings of 'I went by'. Student contributions were taken seriously, there was time for describing and taking different points of view, and there was room for clarification.

There was no explicit contrast between the students' interpretations and the canonical 'mathematical answer'. Instead the teacher clarified and connected different interpretations of students' products. The teacher did not define a correct interpretation of the scales. Instead, the teacher described how she interpreted the two scales using the concept of unitizing. She described how the scales were the same, how they were different, and she described to the students in detail how she interpreted the scales and tick marks on each of the graphs, using the syntax of the representational system, equal intervals and the concept of unitizing.

This analysis suggests that students' multiple interpretations of the scales and meanings for phrases can be used as resources for a mathematical discussion. The teacher supported this mathematical discussion by, rather than evaluating student work, describing in detail how she understood each student's descriptions. The role of this teacher stands in contrast to more traditional roles for teachers in mathematical discussions (Mehan, 1979; Stodolsky, 1988; Thompson *et al.*, 1994). Discussions that make multiple meanings and interpretations explicit and compare different meanings can provide important opportunities for students to appropriate more mathematical or canonical ways of talking.

Multiple interpretations can serve as resources for instruction in bilingual classrooms. A positive perspective on multiple interpretations is particularly important for bilingual classrooms. This mathematical discussion shows that multiple interpretations need not be seen as obstacles but can be used as resources for explaining and using important mathematical concepts such as unitizing. This positive perspective on multiple interpretations shifts the emphasis from asking what difficulties bilingual students encounter to how instruction can support students in participating in discussions.

The role of the teacher in supporting this mathematical discussion is also important to consider for instruction in bilingual classrooms. This example, which transpired in only one language, English, can serve as a model for monolingual teachers working with bilingual students. This teacher supported the mathematical discussion using multiple interpretations, building on students' own views of the scales and grounding her explanations in a mathematical concept. These strategies can serve as a model for engaging bilingual students in discussions that simultaneously connect to student interpretations and keep the discussion mathematical.

Acknowledgements

This work was supported in part by a grant from the National Science Foundation to the author (No. REC-0096065, Mathematical Discourse in Bilingual Settings) and a grant from NSF to the Center for the Mathematics Education of Latinos/as (CEMELA, No. ESI-0424983). Any opinions, findings, and conclusions or recommendations expressed in this material are those of the author and do not necessarily reflect the views of the National Science Foundation. I would like to thank the Chèche Konnen Center at TERC and the classroom teacher.

Note

1. Transcription conventions: [ ] marks the beginning or end of overlapping utterances; = indicates 'latched' utterances that continue without an intervening pause. Timed pauses (1.8) are measured in seconds. A full stop (period) indicates falling pitch or intonation at the conclusion of an utterance; ? indicates rising vocal pitch or intonation at the conclusion of an utterance; ! marks the conclusion of an utterance delivered with emphatic and animated tone. A comma indicates a continuing intonation with slight upward or downward contour. ( ) indicates talk for which transcriber doubt exists. (( )) encloses transcript annotations.

Chapter 7
Reflections on a Medium of Instruction Policy for Mathematics in Malta

MARIE THERESE FARRUGIA

In 1999, the Ministry of Education in Malta published a new National Minimum Curriculum (NMC) for all Maltese primary and secondary schools. The aim of the document was to encourage schools to reflect on issues such as assessment, inclusion, creativity and technology, and to come up with school-based policies on such matters (Ministry of Education, 1999). This document was a step in the direction of school autonomy in a system of education that has been centralised for decades.

One of the points mentioned in the NMC was bilingualism. Although Malta has its own national language, Maltese, English is also widely used as a result of 165 years of British colonial rule that ended when Malta gained independence in 1964. Indeed, Maltese and English are the country's two official languages. Although Maltese is widely spoken as a means of daily communication and is the official language of parliament and the courts, English is essential for international communication, the tourism industry and in local education (Camilleri-Grima, 2003). In the educational context, written texts are more often than not in English, being generally British publications. The writers of the NMC expressed a strong desire that students develop a good knowledge of both languages and suggested that mathematics, science and technology be taught through the medium of English, apparently assuming that this approach would help students enhance their knowledge of the language.

Dilemmas regarding media of instruction are not unique to Malta. Tollefson and Tsui (2004) highlight the fact that, in many multilingual countries around the world, language choice raises a fundamental and complex educational question: what combination of instruction in students' native languages (in our case Maltese) and in a second language of wider communication (in our case, as in many other countries, English) will ensure that students gain both effective subject content education, as well as the second-language skills necessary for higher education and employment?

While not wishing to underestimate the importance of good knowledge of English, I would like to reflect on the implementation of the NMC language recommendation in relation to the teaching of mathematics. In this chapter, I problematise this recommendation by discussing what I consider to be 'tensions' between the recommendation and *other* NMC principles relating to language. I also reflect on the recommendation with respect to its potential impact on pupils' talk in mathematics classrooms. Through these discussions, I hope to highlight how important it is to formulate policies in such a way that they 'sit well' with other educational ideals. Finally, I consider three medium of instruction options for mathematics that are open to us in Malta. My reflections are based on interviews and classroom observations that I carried out in a girls' primary school as part of my doctoral research project (Farrugia, 2007). The aim of the study was to reflect on the teaching and learning of mathematical vocabulary. My interest in language was two-fold: language as a medium of instruction and also as a means to convey meanings for terminology. I will start by briefly describing the empirical context of the study.

The Empirical Context of My Reflections

I chose two classes in order to carry out an in-depth study of the teaching/learning process. The classes were Grade 3 and Grade 6 (7–8 years old and 10–11 years old, respectively). I first interviewed the class teachers about their objectives for a forthcoming set of lessons on a particular mathematical topic such as 'graphs' or 'length'. I then sat in for, and recorded, the lessons, focusing my attention on episodes when mathematical terminology was used or explained (e.g. *multiplication, divide by, measurements, x-axis, regular*). In all, a total of 23 lessons (34 h) were observed consisting of two 'topics' per class.

The lessons were carried out through English, the girls' second language. In both classes, a 'whole-class' teacher-directed approach was used. The teachers, who I call Rose and Gina, tended to teach from the front of the class, guiding the pupils by means of a series of questions. After a topic was finalised, I interviewed the teachers about their own, and the pupils', use of a selection of mathematical words related to that topic; I also interviewed a number of pupils about their understanding of the same words or expressions. During these interviews, we also discussed the use of English as the medium of instruction. For the purpose of this chapter, it is the latter point that is of particular relevance, together with the general language patterns observed during the lessons.

The academic year in which I observed the classrooms (2002–2003) was the first year that the policy had been officially stipulated. English was to be used for all subjects except for social studies, religion and Maltese, and the girls were also expected to address administrative staff in English. The Head of School reported to me that the majority of parents had been in favour of using English as a medium of instruction, while Rose and Gina themselves explained that they believed the approach would help improve the pupils' English and help them understand the textbooks. Both teachers acknowledged that there was still a long way to go, but they were generally optimistic. Rose, however, admitted to me that, based on her previous experience, she believed 'deep down' that the children would understand better if she used more Maltese. Yet she chose to suppress her personal views in favour of the new school policy, saying 'The Curriculum [NMC] tells us to use English more.' Indeed, the Head of School and both teachers considered the NMC document to be a major influence on the school's policy and they seemed to accept the document as an authority that 'must be right'.

It is interesting to note that in the document, the NMC is *personified* through expressions such as 'the NMC encourages' (Ministry of Education, 1999: 79), 'the NMC accepts' (p. 82), the NMC advocates' (p. 79), and so on. I feel that this apparent detachment from human agency masks the fact that the document was written by a group of people, albeit after a period of consultation with various stakeholders between the draft and final versions, all with particular backgrounds and opinions. I believe that a first step in any discussion about a policy is to recognise the human element behind it, and to accept that opinions can be questioned and challenged. In order to contextualise the NMC recommendation, I will briefly explain the general socio-linguistic situation in Malta, as well as how English and Maltese are used in educational settings.

The Use of Maltese and English in Malta

Malta, being geographically positioned in the centre of the Mediterranean Sea, has attracted a succession of powers over the centuries. The Maltese language originated during the Arabic domination (870–1249) as a dialect which, however, went on to develop independently as a language in its own right (Brincat, 2006). Over the centuries, the vocabulary was extended through Romance elements (Camilleri-Grima, 2003). The language also absorbed English vocabulary, first during the period of British colonisation (1800–1964) and more recently, as a result of the influential status of English as a global language. The majority of Maltese people use Maltese as their first language, although many families try to introduce their children to English vocabulary from a young age. Popular perceptions persist in associating English with higher levels of education and social standing: Camilleri (1995) points out that a small portion of the population use English as their home language, and a few private schools promote English as a medium of instruction and for written communication with parents.

The extent to which a Maltese person may use, or attend to, either language will depend on their background, language preference and the context in which they find themselves. In practice, it is difficult to compartmentalise the two languages and many people, including myself, often find themselves interspersing both languages. Indeed, Camilleri-Grima (2003) notes that by 'codeswitching', a person can appear to know enough English to be considered educated, while espousing a Maltese identity. Baker (2001) defines 'codeswitching' as the practice of deliberately alternating between two or more languages. When the switch within a sentence is for only one or a few words, this may be referred to as 'code-mixing', while longer stretches may be considered as 'switches'. As Baker points out, the distinction in practice between mixing and switching is not clear-cut and I will use the expression 'codeswitching' to refer to both.

The two languages are taught formally at school as from the first grade and, at least in principle, English is taught through English, and Maltese through Maltese. Social studies (in primary schools), Maltese history (secondary schools) and the Catholic religion are taught through Maltese, since these subjects are closely tied to the local culture and textbooks in Maltese are available. However, for mathematics and general science in the primary school, and other subjects such as biology and economics in the secondary school, English is the language of written texts, including books and computer software (often UK publications), handouts, whiteboard work, copybook notes and examinations. Therefore, while the spoken language is mainly Maltese, teachers and students shift from one language to another in the course of a lesson; indeed, in her study of various Maltese secondary school subjects, Camilleri (1995) found that linking with written texts was the most common reason for codeswitching.

One situation where this use of codeswitching is particularly noticeable in primary school mathematics classrooms is when the subject at hand is 'story sums' (also known as word problems – see chapters by Halai and by Barwell, this volume). For example a teacher might read out a story sum from a textbook, translating as she goes along as illustrated below. (Note: I use a **bold** print for Maltese or, in the right-hand column, translated speech).

'Jason has twenty copybooks. He gives five to his sister .... OK, **mela Jason għandu xi pitazzi ... għandu għoxrin pitazz. Jagħti ħamsa lil oħtu ...**'	'Jason has twenty copybooks. He gives five to his sister .... OK, **so Jason's got some copybooks ... he's got twenty copybooks ... He gives five to his sister ...**'

Codeswitching in mathematics classrooms may also be the result of stating subject-specific or 'technical' words in English. This form of codeswitching may arise because a Maltese equivalent for these words does not actually exist, as in the case of the mathematical expression *square root*. On the other hand, translations do exist for many words, particularly for mathematical vocabulary used at primary school level. However, it is more common to use the English versions as part of the 'academic' classroom talk, rather than the Maltese ones. The example below illustrates this point: all the English words used by the teacher in this excerpt may in fact be substituted by Maltese equivalents:

'**Ħa npinġu** square ... four sides,   '**Let's draw a** square ... four sides,
   kollha indaqs, u erba' angles **ta'**   **all equal, and four** angles **of**
   ninety degrees'   ninety degrees'.

Although codeswitching in many classrooms is a common practice in Malta, the writers of the NMC appeared not to agree with it. Hence, they suggested that mathematics, science and technology at primary level, and other subjects such as biology and economics at secondary level, be taught through English. Codeswitching was only acceptable when the use of English caused 'great pedagogical problems' (Ministry of Education, 1999: 82). I cannot exclude the fact that some of the writers may have had a personal perception of English as being in some way 'superior' to Maltese, or, as Baker (2001) implies, that the recommendation is motivated by an ulterior agenda to offer an advantage to certain social groups. However, from my personal acquaintance with some of the writers, and my interpretation of the document, the *apparent* reasons for the recommendation were to find a way to improve students' competence in English and a disapproval of codeswitching as a pattern of language. The medium of instruction issue for mathematics is a hotly debated topic in Malta. Those in favour of English argue in a similar way to the NMC writers, while those in favour of Maltese (or rather codeswitching) tend to present arguments that prioritise mathematical understanding.

The medium of instruction debate is not a new one to Malta. As in other countries, such debates have always been shaped by political, social and economic forces (see Tollefson & Tsui, 2004, for a comprehensive compilation of such stories). For example the use of *Italian* as a medium of instruction in the mid-19th to early 20th century in British-colonised Malta reflected the preference of the professional elite and the nobility. However, in 1879, a report on the educational situation in Malta commissioned by the British government suggested that *Maltese* be used as a medium (Keenan, 1879) and in the subsequent decades there was much 'to-ing and fro-ing' between various official published reports or pressure by interested parties for one of the three languages. Generally, Italian was preferred by the educated elite, but finally fell out of favour in the 1930s due to its association with fascism; English increasingly gained relevance for the civil service and military economy of the period; and Maltese was the language of the people, strongly promoted by a group of distinguished Maltese writers in the 1920s and established as the national language in 1934. Still, Camilleri (1995) states that English tended to be preferred by school authorities in the 20th century because of its importance as the colonial language and possibly because teachers were trained by British religious orders until the late 1970s, when the responsibility was moved to the University of Malta.

Today, it appears that teachers, especially those in state schools, tend to prefer Maltese or codeswitching, since, as explained by Camilleri (1995), this allows a flexible and comfortable mode of communication and hence serves as a useful pedagogical and communicative resource. Camilleri makes this point with respect to a variety of subjects. When considered more specifically for mathematics, her conclusion is consistent with Setati and Adler's (2000) view that, in the South African mathematics classrooms they observed, codeswitching between the school language, English, and Setswana appeared to be something positive, in that it enabled learners to 'harness their main language as a learning resource' (Setati & Adler, 2000: 244).

The Teaching and Learning of a Subject Through a Second Language

The method of targeting a second language by teaching all or part of the curriculum through this language is known as the immersion approach (Baker, 2001). It originated in Canada in the 1960s, the success of which Baker attributed to teacher enthusiasm and competence in both English and French, together with parental motivation and support. There are different forms of immersion. Baker lists 'early' (i.e. at a very young age), 'delayed' (starting at 9–10 years old) and 'late' (secondary level) immersion. The *time* of immersion may also be varied, from 'total' (starting with 100% immersion at a young age, reducing to 80% and finally 50% by the end of junior school) and 'partial' immersion (i.e. 50% through infant and junior school). A comparable idea that has become popular in the European Union in recent years is Content and Language Integrated Learning or CLIL, whereby a subject like mathematics is taught through a targeted foreign language. In such contexts, teachers are trained in both the subject and the language. Within the NMC document, it is not evident on which of these approaches the writers draw.

My initial reaction to the NMC recommendation for English was to agree with it, being influenced by my own experience of using English frequently in personal and academic situations, and having experienced all my schooling through English in a private school in the 1970s. However, conversations with a linguist encouraged me to reflect beyond my own particular experiences and to recognise the issue as a complex one. This in turn prompted me to realise that those involved in establishing policies related to teaching multilingual students need to reflect on and discuss assumptions and beliefs regarding language. I came to appreciate how important it is to recognise linguistic bias and consider colleagues and students who have different linguistic backgrounds and experiences, and hence, different opinions and feelings about language use in the mathematics classroom.

In the sections that follow, I will discuss the recommendation regarding the use of English as a medium for mathematics *vis-à-vis* other NMC principles. I will also reflect on the promotion of pupils' verbal participation in class, both generally and in terms of mathematical language.

The Principles of 'Consistent Use of Language' and 'Pedagogy of Cooperation'

While the writers of the NMC document stated that codeswitching could be used for mathematics in the case of difficulties in communication, they also present the following recommendation:

> The National Minimum Curriculum advocates consistency in the use of language during the teaching-learning process. (Ministry of Education, 1999: 79)

My own interpretation of this statement is that teachers are to use their discretion as to whether to use Maltese or English, and to avoid 'unnecessary' codeswitching. Immediately, I am faced with the question of what is 'necessary' or otherwise, and I realise that this is something subjective and context-bound to be decided on by the individual teacher. However, it is interesting to note that the teachers I observed appeared to interpret the statement to imply the continuous use of English. Perhaps this assumption was based on the promotion of English at other points in the NMC document, and on the fact that attempting to use *Maltese* 'consistently' would create difficulties because of the local practice of retaining mathematical words in English; hence by default, the consistent language is necessarily English. These different readings of the NMC statement illustrate how a policy statement is open to multiple interpretations.

Using English 'consistently' seemed achievable in the Grade 3 classroom, thanks to the style of interaction promoted by the teacher, which was almost exclusively of the 'Initiation-Response-Feedback' type (Sinclair & Coulthard, 1975). That is pupils tended to give short responses to the teacher's questions, which were then confirmed or otherwise by the teacher. For example:

(The class is correcting a textbook multiplication exercise. Given a number of gloves, the pupils were required to find the number of fingers).

Teacher:	Kelly, how many gloves did we have? [INITIATION]
Kelly:	Two. [RESPONSE]
Teacher:	Two. And how many fingers? [FEEDBACK (CONFIRMATION)] then [INITIATION]
Kelly:	Five. [RESPONSE]
Teacher:	And that gives me? [IMPLIED ACCEPTANCE then INITIATION]
Kelly:	Ten. [RESPONSE]
Teacher:	Ten. [FEEDBACK]

Thus, the girls gave only short answers, often consisting of mathematical words which, in Malta, are often retained in English anyway. On the other hand, *pupil-to-pupil* talk took place in Maltese. This may have been social talk as, for example, when a pupil asked a classmate about her broken arm in plaster. Other instances when Maltese was used were during the two occasions that a group activity was carried out. For example:

(The girls are working in groups of four, estimating and measuring various objects).

Sandy:	Kemm taħseb li hi? ... Naħseb **forty**.
	[**How much do you think it is? ... I think it's** forty]
(...)	
Roberta:	Ikteb x'taħseb, imbagħad kejjel.
	[**Write down what you think, then measure**].

Hence, while the whole-class interaction offered consistent use of English, pupils' interaction did not. This is a similar situation to that observed by Setati and Adler (2000) in South Africa. In the contexts observed by these researchers, pupils used their first language interspersed with mathematical English for group interaction, but switched to English for the 'public' domain, that is, the whole-class interaction. Interestingly, the NMC document recommends what it calls a 'new pedagogy of co-operation' (Ministry of Education, 1999: 35) that promotes discursive and collaborative group work, rather than the prevalent competitive and individualist approaches. I welcome this recommendation, but it may be the case that primary school pupils might communicate more comfortably and effectively in their first language than in their second when working in groups. Hence, the recommendation for English may not sit well with the promotion of cooperative work.

In Grade 6, Gina did codeswitch a couple of times during the topic 'Graphs' in order to help the girls tackle the task at hand. As she admitted, 'One thing I always make sure is that I don't sacrifice a maths lesson for English.' Thus, Gina was experiencing what Adler (2001) referred to as a dilemma of codeswitching, that is, finding the balance between using the first language to aid understanding, and using English to provide access to a language deemed useful not only for mathematics, but also for other contexts. Gina also sometimes used Maltese when she interacted with pupils on a one-to-one basis as she did when she walked around the classroom monitoring pupils' written work. However, she generally tried to stick to the school policy as much as possible, and hence I could say that *generally* she was consistent in her use of English.

I noted more Maltese used by the Grade 6 *pupils* than I had noted in Grade 3. First of all, more pupil-to-pupil talk occurred, since the girls were allowed to talk to each other quietly as they worked on an exercise. In these contexts, the girls used Maltese. The pupils also contributed more to classroom interaction than their Grade 3 counterparts, and frequently used Maltese to do so. In such instances, Gina might ask the pupil to repeat in English, as in the following excerpt:

(*The teacher has started drawing a graph on the whiteboard. She has written two sets of scales in the course of her explanation*).

Kirsty: Miss, għaliex għandna tnejn [skali]?
[Miss, why do we have two [scales]?]
Teacher: English!
Kirsty: Miss, why do we have two, em, x-axis and y-axis | (sic)?
Teacher: Because first I explained, then I had to draw. Then I explained again! They're both the same.

However, there were many occasions when Gina did *not* insist the pupils switch to English. Rather, she accepted the contribution but answered in English herself. Hence, the language used in Grade 6 was not 'consistent' in the way the NMC writers presumably intended. I think that if a teacher allows pupils to use Maltese, this creates a tension between the implementation of the English language and consistency recommendations, and undermines the whole objective of using English. After all, the main problem perceived in Malta is not actually *exposure* to English, but production (speaking and writing) of the language. Furthermore, I believe that consistency of language use may be generally easier for a *teacher* to achieve than for the pupil, assuming that a teacher feels confident using English. It is worth mentioning that while the teachers I observed were fluent in English, other teachers may lack this confidence and flexibility of language, so that they themselves may not be in an ideal position to lead class talk through their second language.

If, by 'consistency of language', the NMC recommendation had indeed implied the use of English throughout, then the writers of the NMC may have underestimated the difficulty encountered in insisting that pupils use their second language to discuss at length. Furthermore, they may have inadvertently assumed that, in a mathematics classroom, it is the teacher who is to do most of the talking. While this may have been the case traditionally, today it is considered much more desirable for pupils to contribute more significantly to classroom talk. I will take up this point in more detail shortly, but will first reflect on another NMC ideal, that of inclusion.

The Principle of 'Inclusion'

One educational aspect that is given prominence in the NMC is the idea of inclusion, expressed as follows:

> An inclusive education is based on a commitment ... to fully acknowledge individual difference... (Ministry of Education, 1999: 30)

One difference that may be present in a Maltese classroom is pupils' varying levels of proficiency in English. Such variation may be a consequence of the degree of exposure to, and support for, the language outside school which, in Malta, tends to be higher in middle class homes. Indeed, I believe that the intention of the writers of the NMC is precisely to narrow the gap between the pupils' different levels of proficiency. They appeared to believe that this can be achieved through further exposure to English, a belief echoed in Rose's comment, 'What they lack at home, we make up here'.

It is interesting to contrast the general satisfaction expressed by Rose and Gina regarding the approach being used, with the opinions expressed by the pupils interviewed. I asked the pupils for their opinion regarding the use of English as a medium of instruction for mathematics. There was a tendency for those who sounded more confident to be the pupils that their teachers described as the 'higher achievers' in mathematics, while what they referred to as the 'average' and 'weak' pupils were more likely to express some reservations related to potential understanding. For example:

> Fiona, Grade 3: 'I don't know too much English ... Sometimes I understand the lesson, sometimes I don't ... I don't like asking (...) [when I do ask] I ask in Maltese because I'm afraid to include something in English.'

The tendency to prefer Maltese was more evident with the younger pupils. One reason for this may have been that, by virtue of their age, the girls may generally have had less experience with English than the older girls. Another reason may have been that the Grade 3 teacher tended to adhere to the language policy more strictly than the Grade 6 teacher, offering the pupils less flexibility with language.

Another indicator of language preference came through the interviews themselves. For the purpose of another aspect of my study, part of the pupil interviews was conducted in English. Three of the five pupils described as 'weak' pupils by their respective teachers pulled faces and initially showed a reluctance to speak; they switched back to Maltese at the first opportunity. Not all the 24 pupils I interviewed expressed reservations, however. Some of them said it made no difference to them in which language the lessons were conducted, while two even said that they preferred English to Maltese. However, I suggest that as part of the local debate on the medium of instruction for mathematics, it would be useful to consider the discomfort and possible lack of understanding that pupils may experience when required to use their second language. If a practice of inclusion is considered to be a desirable one by the NMC writers, then I suggest that further reflection may be necessary on how language can be used in an inclusive way. One possible alternative may be to teach through codeswitching while offering *explicit attention* to mathematical English statements that are likely to be met in writing. This would concur with researchers' recommendations to give explicit attention to mathematical language, especially when it is presented in a second language (e.g. Rothman & Cohen, 1989; Zaskis, 2000). The teachers I observed seemed to assume that the pupils would 'pick up' English, including mathematical English, in time.

I conjecture that explicit attention to language may be a more effective way to help 'close the gap' between pupil differences in English as the NMC writers desire. Here again is an example of how a policy statement may be interpreted differently: my own beliefs on how equity can be achieved differ from those implied in the NMC document, and from those apparently held by the teachers I observed. Thus, the inclusion principle may be worth reflecting on as part of our local discussions on the use of English.

Promoting Pupils' Verbal Participation

The importance of increasing pupils' contribution to classroom talk was brought to the fore in the UK and also in Malta, through the *The Cockcroft Report* (DES, 1982). This stated that students should be encouraged to discuss and explain the mathematics they were learning. Indeed, pupil talk is useful in the mathematics classroom, since it enables understandings to be clarified and misconceptions to be addressed (Griffiths & Clyne, 1994). Furthermore, Pimm (1987) recognised pupil discussion as a means for talking things through and organising one's thoughts. Given the importance attached to pupil talk, I wished to explore to what extent the pupils I observed were able to contribute to classroom interaction, given that they were expected to use English.

As already explained, pupils' contributions in Grade 3 were limited by the style of interaction. I believe that this was Rose's preferred pedagogical style, but I cannot exclude the fact that she used this style because she was conscious of the potential difficulty the pupils might find in using English. On the other hand, although a similar question-and-answer style was often used in the Grade 6 class, Gina also encouraged her pupils to use more speech by asking open questions or inviting explanations. Indeed, according to Clemson and Clemson (1994), comments and questions like 'Tell John how it works', 'Go on ...', 'Where did the eight come from?' are useful to promote pupil talk. I noted that at such points in the lesson, however, the Grade 6 girls often showed some difficulty in expressing themselves, and resorted to using gestures. For example:

(A class discussion is taking place regarding methods primitive man may have utilised to measure objects).
Teacher: Can you use your body to measure?
Monica: Yes!
Teacher: How Monica? Tell us.
Monica: *(Shows up her right pointer finger).*
Teacher: Your finger. Tell me how. Measure the desk with your finger.
Monica: *(Lays her finger along the edge of the desk and moves it along like a worm. The other girls laugh).*
Teacher: That's a bit difficult, but good!

Hence, although invited to tell, Monica *showed* instead. Of course, gestures are an integral part of any communication and, indeed, both the teachers themselves used gestures as they taught. However, in the teachers' cases, gestures tended to *complement* the speech, such as by pointing, rather than *replace* it.

Sometimes Gina would rephrase a pupil's contribution or 'fill in' the missing language, as in the following excerpt:

(The pupils are working on a tiling problem. The teacher has just distinguished between area and perimeter. Dorianne looks up at the ceiling beam that runs across the width of the room).
Dorianne: That ... *(points to the beam).*
Teacher: The beam.
Dorianne: *(Shrugs her shoulders as though to say 'whatever').* Em, it's like perimeter, it's around. But if it's like a wall ... It's the area *(moves hand up and down as one would do when painting a wall).*
Teacher: Ah, we can take the length of the beam, she says, but if we need to wall the space in between, we have to find the area.
Dorianne: *(Nods).*

This is not to say that the Grade 6 pupils found difficulty with every contribution. There were many occasions when they offered clear responses of say, one or two sentences long. For example:

(The class is reading off information from a graph).
Teacher: Why do the numbers in the vertical axis stop at hundred? Caroline?
Caroline: Because the total [highest] mark is hundred.

Most of the Grade 6 pupils indicated that they could express themselves effectively in statements of this length, although the accuracy of expression (in the sense of correct use of English) varied from pupil to pupil. However, generally speaking, the longer the response expected, the greater the tendency for the girls to falter, express themselves vaguely or use gestures. This type of contribution was accepted by Gina; she explained to me that she was reluctant to discourage her pupils' efforts to use English. I concluded that although Gina generally tried to encourage her pupils to express themselves by asking open-ended questions, using English appeared to be restricting the pupils' contributions. Thus, a tension was created between the ideal for using English and the promotion of pupil talk.

The Use and Development of Mathematical Language

The talk used in a mathematics classroom is likely to consist of a mix of 'everyday' and 'mathematical' vocabulary such as *shape*, *graph*, *twenty*, *multiplication*, *length* and so on. According to Harvey (1982), 'technical' language is not always essential and pupils may well use informal language to express themselves. For example a child might call an *angle* a *corner*, or refer to the *perimeter* of a shape as the *outside line*. However, Harvey also states that more technical language is convenient, since standard words or expressions increase the potential of more effective communication with others in, and beyond, the immediate classroom. Hence, it is useful that teachers help pupils to use more conventional language (Miller, 1993) which, according to Pimm (1995), allows us to talk about things and to 'point' with words. Learning to speak mathematically implies learning to *mean* mathematically (Pimm, 1987) and hence, when pupils have been taught to use mathematical language to express their ideas, the teacher no longer has to guess at the state of pupils' learning but can act to extend that learning (Lee, 2006).

Mathematics, like other disciplines, has its own set of expressions and ways of saying that collectively can be referred to as a 'mathematics register'. A register was explained by Halliday (1978) as words and meanings that are specific to a particular *function* of language, in our case, the communication and expression of mathematical ideas within a classroom setting. This, of course, includes the vocabulary related to the discipline such as *subtract*, *graph*, *square root* and so on. However, Pimm (1987) also draws attention to the use of argumentative phrases ('if and only if'), referencing conventions ('Square ABCD') and the use of metaphors ('the slope of a graph'). Furthermore, with respect to the *written* register (which tends to be more formal than the spoken one), Morgan (1998) suggests that the following render a text mathematical: the use of imperatives ('drop a perpendicular'), the passive structure ('a line is drawn'), nominalisation (i.e. changing a process – *rotate* – into a noun – *rotation*), diagrams and tables (see also Monaghan, this volume).

It seems to me that three options for register use are available to us in Malta, namely: using oral and written English registers; using an oral codeswitching register and a written English one; or using oral and written Maltese registers. The first is the option that Rose and Gina were attempting and I have outlined some difficulties that arose in that situation.

The second option is that which several teachers continue to believe to be a helpful method, given that publications will remain in English for the foreseeable future. However, if we accept that oral codeswitching is a useful approach, I would like to once more stress that teachers can make more explicit the move from the informal spoken 'mixed' register to the more formal written English register (see Clarkson, this volume). Indeed, Morgan (2007) points out that if a teacher stuck only to informal language, in our case Maltese/codeswitching, some students would not be provided with 'access to higher status forms of language' (p. 241). Morgan poses the challenge of how teachers can coordinate the everyday and the specialised languages in order to facilitate learning for all. I consider this challenge to be one that we should certainly take up in Malta. As a starting point for reflection, I might offer an illustration of a strategy I observed being used by a teacher called Angela during the teaching of the topic 'Money' in a Grade 3 class. As the class focused on shopping, prices, cost, change and so on, Angela shifted from Maltese to English consciously, formulating the same questions and statements in both languages especially prior to any written work. She pointed out translations for the common Maltese words explicitly, while progressively using the English versions more as the topic developed. She also encouraged her eight-year-old pupils to express ideas in simple English. For example when learning about change, the pupils were encouraged to suggest an 'ending' to a shopping transaction, by offering statements such as 'John has 30 cents change', 'The change is 30 cents' and so on. The teacher used her discretion regarding if and when to correct any grammatical mistakes. This approach appeared to contribute not only to the understanding and the development of 'mathematical' vocabulary, but also to the development of general English competence.

The third option available to us is to use Maltese for both spoken and written mathematics. This would present the challenge of somehow encouraging the use of Maltese mathematical words in classrooms if these already exist. The same holds for those words that can be found in a good dictionary (e.g. **assi** for *axis*) but which, from my experience, I can say are rarely used as part of the classroom register. On the other hand, if a word or expression does *not* exist, we need to agree on how to express it in Maltese. It is not easy to define in a clear-cut manner which words might have Maltese translations. As a very general rule, words that have common everyday usage such as *length, height, coin, add, subtract* and so on are more likely to have commonly used Maltese equivalents than terms that I might loosely call more 'technical' such as *axis, polygon, drop a perpendicular*.

Unlike mathematical English which 'exists', we cannot claim to have an established spoken and written Maltese mathematics register. This does not mean that such a register is not possible. The Welsh, Māori and Aboriginal Australian communities have gone through the trouble of developing mathematical language in their own languages (see, respectively, Barton *et al.*, 1998; Jones, 1998; Roberts, 1998). The exercise of developing a Maltese mathematics register could be undertaken by local mathematics educators and linguists. A standard Maltese register would then allow us to produce written texts in Maltese and eliminate the need for codeswitching. However, at present this remains only an interesting possibility, one that might seem overwhelming because of the financial implications of the project. It is worth noting, however, that such an endeavor would, in fact, fulfill yet another NMC principle, that of 'strengthening of the Maltese language' (Ministry of Education, 1999: 37). I feel that using English for mathematics works *against* this ideal since the method may actually pass an indirect message to young learners that Maltese is not a suitable language for mathematics, thus detracting from the value of the language. Furthermore, when English is used as a medium for mathematics, Maltese mathematical vocabulary is neither used nor developed explicitly.

Conclusion

The Maltese NMC contains several principles that I welcome as a mathematics educator. However, from the observations I made in two primary school classrooms, I concluded that the recommendation for using English, our second language, for mathematics, appears to work counter to these principles. In particular, in this chapter, I have reflected on the ideals of consistency of language, cooperative learning, inclusion and the strengthening of the Maltese language. Using English for mathematics also appears to work counter to the belief held in contemporary mathematics education that increased pupil talk is something desirable.

I have generally argued against the use of English, and viewed codeswitching as potentially useful, rather than as a problem. Teachers may very well have developed strategies over time to help them cope with the local situation in which English texts are used, but an official recommendation by policy makers in favour of using English appears to undervalue these strategies. Indeed, the NMC suggestion may actually reduce the need to identify and share such strategies with others. Rather than trying to eliminate this situation, I suggest that we look for ways to maximise the effectiveness of linking strategies. I also believe that the notion of mathematical language or register is an aspect of the subject that must not be overlooked as we engage in our local medium of instruction debate. I have suggested three options regarding spoken and written registers that are available to us, each needing careful consideration.

It is worth noting that, while I have rooted my discussion in the mathematics classroom and, in particular, focused on the verbal aspect of the subject, some wider implications can be drawn. My reflections illustrate how practitioners may hold different interpretations of the same policy statement; they also highlight that a recommendation cannot be considered in isolation, but rather must be evaluated in relation to other principles. Hence, I believe that the key to exploring effective ways to use language in the mathematics classroom lies in dialogue. Living in a small country (a population of 400,000 living on 316 km^2!) means that it is relatively easy to find opportunities to hold discussions, even informal ones, with policy makers, academics and teachers. Indeed, through dialogue, the original NMC statement published in the *draft* version of the document, which had read '**the NMC obliges the teachers in this [primary] sector** ...' (Ministry of Education and National Culture, 1998: 21, translation mine), was toned down after an interim consultation period to read: 'the NMC encourages teachers ...' (Ministry of Education, 1999: 79).

More recently, again after discussion that was possible as part of a National Conference for Mathematics, the Ministry of Education published another document that reported on the conference and also included recommendations formulated by an Action Group of which I was a member. With respect to language use, we recommended that the language used should be one that helps children understand mathematical concepts. We also suggested that the use of English for mathematics should not be considered in isolation, but should be part of a whole school policy for bilingualism. Finally, we advised that the language policy needs to be implemented on a school and class level (Ministry of Education, Youth and Employment, 2005).

Continued discussion and reflection will ensure that, in Malta, we continue to strive to find the best way to use language for the benefit of our students. Ultimately, I think that we all agree on one fundamental principle: that the language used and developed in the mathematics classroom should offer our children the best opportunities to participate actively and to grow to understand and appreciate the subject.

Chapter 8
Bilingual Mathematics Classrooms in Wales

DYLAN V. JONES

With a population of around 3 million, Wales is one of the smallest of the four nations that make up the United Kingdom. Although certain, mainly small, pockets of Wales are culturally and linguistically diverse, the vast majority of the population are fluent English speakers. Welsh, the native language of Wales, is spoken by just over half a million people and nearly all of these Welsh speakers also speak English. Welsh does not have the same official status as English although The Welsh Language Act 1993 aimed to ensure that Welsh and English are treated equally within certain contexts such as in public services and in the courts. Over recent years, however, there has been a revival in the fortunes of the Welsh language and this revival has been most noticeable within the Welsh-medium and bilingual education sector. In common with other contexts described within this volume, much of what can be observed within bilingual mathematics classrooms in Wales must be considered within the cultural, political and social context within which teaching and learning occurs.

Until the middle of the last century, English was largely unchallenged in Wales as 'the language to get on in the world' and was almost universally used in state schools. In response to increasing anglicisation, however, resistance, particularly from a small number of Welsh intellectuals, spawned a number of political and cultural movements that contributed to a reversal in the fortunes of the Welsh language. At the end of the 19th century, Welsh was spoken by the majority of a population of around 1.5 million but by 1991, the number of Welsh speakers had declined to just over half a million or just over 18.7% of the population. Significantly, however, the 1991 census revealed an increase in the number of young people who spoke Welsh and by 2001 census figures showed that 20.8% of the population of Wales could speak Welsh (Welsh Language Board, 2003).

While many factors have helped to achieve what has been described as 'a truly remarkable cultural achievement' (Aitchison & Carter, 2004), there is no doubt that one of the most important contributors to the language's renaissance over recent years has been the growth of Welsh-medium and bilingual education (Jones & Martin Jones, 2004). Little more than half a century after establishing the first Welsh-medium primary school, Welsh is now the sole or main medium of instruction in 458 (29%) of all primary schools in Wales and 54 (24%) of all secondary schools are classified as Welsh speaking[1] (Welsh Assembly Government, 2007a, 2007b).

Bilingual Mathematics Classrooms

In the more anglicised areas of the north-east and south-east of Wales, most primary and secondary schools are English medium and teach Welsh mainly as a second language via timetabled Welsh lessons. In such schools, few, if any, teachers or pupils may be first-language Welsh speakers and pupils attending these schools often leave school at 16+ or 18+ with little more than a basic grasp of Welsh. In these schools, mathematics, like all other aspects of the curriculum, apart from Welsh as a subject, is taught in English.

In many of these more anglicised regions of Wales, however, designated bilingual schools, *Ysgolion Cymraeg* (Welsh Schools), have been established to offer a Welsh medium alternative to the other local schools. At the primary level, as in many other places such as Catalunya (Catalonia) and Euskal Herria (The Basque Country) these schools mirror the kind of provision found in the Canadian programmes of early immersion (Baker, 1996). Up to 95% of the pupils attending these schools may come from English-speaking homes but, by the time they enter what is likely to be a Welsh-medium secondary school, the vast majority are reasonably fluent in both English and Welsh. Building on what has been accomplished at the primary level, *Ysgolion Cymraeg* at the secondary level go on to teach most, if not every, aspect of the curriculum through the medium of Welsh and, pedagogically, the challenges facing teachers and learners in these schools are similar to those found in other monolingual contexts. Interestingly, however, while one of the arguments used for the development of Welsh-medium mathematics education was the right of pupils to be taught in their native tongue, very many pupils learning mathematics through the medium of Welsh today do so through their second language.

It is a third type of school that presents the particular set of circumstances and challenges that this chapter will go on to examine. In some rural areas, what may be termed *traditional* (or natural) bilingual schools cater for all pupils, whatever their linguistic backgrounds or preferences. Historically, these schools taught most subjects through the medium of English to all pupils but over recent years have increasingly endeavoured to offer more curriculum areas in Welsh as well as in English. In some Welsh heartland areas in the north and south-west of Wales, up to 80% of the pupils attending these traditional bilingual schools may be native Welsh speakers. Like most people in Wales, however, they will also be fluent in English. Other traditional bilingual schools will serve areas where far fewer pupils are first-language Welsh speakers. Whatever the proportion of first-language Welsh speakers, traditional bilingual schools must also cater for non-Welsh-speaking pupils who may be local or who may have recently moved to the school's catchment area from neighbouring England or elsewhere. The relative size of the various linguistic groups within any individual school impacts upon the choices available to school managers, as considered in the next section. While all these traditional bilingual schools will aim to ensure that as many pupils as possible are fluent in both Welsh and English by the time they leave secondary school, for some pupils, this goal is unlikely to be achieved. Pupils who have recently moved to Wales, or who come from English-speaking homes where there may be little support for them to learn Welsh, may demand to be taught in English and the school must cater for them.

It is within these schools that we find two languages explicitly at work in mathematics classrooms and in which practices and issues arise that resonate with those described elsewhere within this book. Much of the research undertaken internationally in bilingual mathematics classrooms has been in contexts where, because of their limited proficiency in English, learners also draw on their native language, in some cases with support from the teacher. In such contexts, drawing on students' first language can help learners learn mathematics and develop their English, as in Adler's (2001) accounts of working with multilingual pupils in South Africa or Moschkovich's (this volume; 1999a) research in Spanish–English bilingual mathematical classrooms in the United States (see also Farrugia, this volume). This chapter discusses practice within classrooms that are at least as complex as those described by Adler or Moschkovich. For some pupils in these classrooms, English is their first language and Welsh their second. For others, the reverse is true. Some will learn mathematics in English and others in Welsh and for some their chosen language for learning will be their first language; for others it will be their second. In many respects, the above combinations possess considerable symmetry. What adds to the complexity, however, is the fact that while all the learners are more or less fluent in English, individual pupils' understanding of Welsh may fall anywhere on the continuum between very limited understanding and complete fluency. Because of this, communications made in English are accessible to all, while those made in Welsh may not be fully understood by everybody. Although sympathetic and supportive of their school's overall aim of developing pupils' bilingualism, mathematics teachers are likely to see the teaching of mathematics as their primary duty and, given the context described above, their task is often one of 'coping with' rather than 'developing' their pupils' bilingualism.

This chapter considers a number of issues that arise in relation to teaching and learning mathematics within a particular type of secondary (11–18) school in Wales where:

- all the pupils and teachers are fluent in English, whether it is their first or second language;
- some pupils and teachers are first-language Welsh speakers, others are competent second-language Welsh speakers, while others have very little Welsh;
- all pupils have expressed a preference or have been guided to learn mathematics in either Welsh or English; their chosen language may be their first or second language.

One such school, which I shall call 'School X', fits the above description and contributed to elements of the research reported below. It is located in a rural part of Wales where about 50% of the pupils are first-language Welsh speakers. The remainder has a proficiency in Welsh, which ranges from being excellent second-language learners to recent immigrants who have very little Welsh. About 40% of the pupils have chosen to study mathematics and science through the medium of Welsh. Although the three sections that follow tackle quite different themes, they all relate to practice within School X and other similar bilingual schools.

The Organisation of Learning

Grouping pupils

Teachers and senior managers in School X, as in many other traditional bilingual secondary schools, are faced with difficult choices in relation to the way mathematics classes are formed. One option is to group pupils according to their linguistic competence or choice, while an alternative possibility is to group pupils according to their aptitude for mathematics. A third option is to create classes that are 'mixed' both in terms of the pupils' ability in mathematics and their linguistic backgrounds or preferences.

The first option leads to the formation of bands (or streams) based on language preferences or backgrounds. Setting may then occur within these bands but individual classes may be quite mixed in terms of the pupils' aptitude for mathematics. The second option is one of setting by ability right across the year group. This approach leads to the formation of classes that contain pupils with a similar aptitude for mathematics but every class will be 'mixed' in terms of the pupils' linguistic preferences or background. A third option is to form classes that are 'mixed' both linguistically and academically. Anecdotal evidence, however, suggests that

---

>
> Consider a traditional bilingual secondary school with an intake of around 110 pupils at 11+. There are two favoured options in relation to the way teaching groups are formed for learning and teaching mathematics
>
> **Option One: Setting within language bands**
> Initially, Welsh-medium and English-medium bands (or streams) are formed. Sets based on attainment are then formed within these bands. Bands may be of a similar size or one may be significantly larger than the other. Below are three examples of the way teaching groups may be formed, depending on the number of pupils in each band. Learning and teaching within individual classes will be in one language only.
>
> | *Welsh-medium band* | *English-medium band* |
> |---|---|
> | Scenario 1: a similar number of pupils in each band | |
> | Set 1   Set 2 | Set 1   Set 2 |
> | Scenario 2: a small number of Welsh-medium pupils | |
> | Welsh class | Set 1   Set 2   Set 3 |
> | Scenario 3: a small number of English-medium pupils | |
> | Set 1   Set 2   Set 3 | English class |
>
> **Option 2: Setting across year groups and the formation of bilingual classes**
> Pupils are grouped solely on the basis of mathematical attainment. This may lead to 4 mathematical sets, with each set containing pupils whose grasp of Welsh may range from very little (if at all) to first language Welsh speakers. Some pupils within each set will have elected to learn mathematics in Welsh whilst others have chosen to do so in English. All pupils are fluent in English.
>
> Set 1   Set 2   Set 3   Set 4

Figure 8.1 The formation of teaching groups in a traditional bilingual secondary school

this is not an option that is generally pursued. Figure 8.1 clarifies the first two options by considering some of the possible grouping alternatives for a hypothetical but typical secondary school that attracts around 110 pupils per school year.

School X, like several other bilingual schools, groups pupils according to linguistic criteria during the first three years of their secondary schooling (i.e. Option 1) but sets pupils according to their attainment in mathematics (i.e. Option 2) from their fourth year onwards in preparation for when they will sit national external examinations at 16+. This pattern of shifting from an Option 1 arrangement to Option 2 after the first year or so of secondary schooling is a model that several schools follow for teaching mathematics because, as pupils approach the external examinations at 16+, teachers generally consider it preferable to teach classes that are linguistically rather than academically diverse.

Structuring lessons

In situations where bilingual learners are grouped together in the same place at the same time with the same teacher, as described by Option 2 above, certain teaching strategies may be more productive than others. When addressing the whole class, teachers will often reiterate key points in both languages. The extent to which such repetition is used by individual teachers varies a great deal and is dependent on a whole range of factors including the linguistic make-up of the class and the teachers' established pattern of language use. Although repeating key points in two languages may provide some pupils with helpful opportunities to understand the mathematics being discussed, a heavy reliance on such a technique, often referred to as a 'concurrent' approach, can also have its disadvantages. As Baker puts it:

> ... the replication and duplication of content is a danger in bilingual classroom methodology. Where the same subject matter is repeated in a different language, some students will not concentrate or go 'off task'. (Baker, 1996: 240)

Drawing on his observations of classes where Spanish and English were used by teachers, Faltis also points to a similar potential for 'switching off':

> Not only do students tune out when the teacher is talking in their second language, they learn quickly that the teacher will translate the talk into their primary language to make sure that they comprehend the language, but not necessarily the concepts, principles, and processes associated with the knowledge system. (Faltis, 1996: 31)

When faced with the type of grouping options described in Option 2, teachers find that they have little alternative other than to spend at least part of their time switching between languages in a fairly systematic manner. If such practice is potentially unhelpful, then it would seem logical to try to reduce the frequency and duration of such episodes. Williams (1995) described, with reference to a higher education context, a particular structure that might be helpful in such circumstances. As can be seen from Table 8.1, it is based around the principle of minimising the amount of time when the teacher needs to address the whole class.

Table 8.1 A possible approach to a concurrent bilingual lesson

Period (min)	Welsh speakers	Non-Welsh speakers	Teacher
5–10	Listen	Listen	Bilingual or English introduction
20	Teacher introduction in Welsh	Group/individual work	With the Welsh speakers
20	Group/individual work	Teacher introduction in English	With the non-Welsh speakers
10–20	Group/individual work	Group/individual work	Circulating: individuals' chosen language
10–20	Respond – Welsh/English	Respond – English	Bilingual feedback

Source: Adapted from Williams (1995)

The design and use of bilingual resources

Teachers working in bilingual contexts, as described in Option 2, above, find that they must prepare resources, for example, worksheets or handouts, either separately in both languages or bilingually, that is, using both languages on a resource that is made available to all learners.

In Jones (2000), I provided examples, including some from School X, which suggested that the choices made by teachers were often influenced by the amount of text that might be necessary on any particular resource. For example if a worksheet contained mainly diagrams or algebraic expressions, it might be more economical and efficient to prepare a bilingual resource on which a limited amount of text is included in both languages. In other situations, such as a worksheet that includes number problems in words, it might be more economical to have separate language resources that are distributed according to learners' language choices.

Educationally, there are a number of issues surrounding the availability or otherwise of bilingual mathematics resources for learners who have chosen to learn in one language. To what extent does having access to both language versions of a resource contribute to learning? Does encouraging pupils to draw on the range of linguistic resources available to them help or hinder their learning? Does switching between different language versions of a resource contribute to a better understanding of the mathematics or restrict their ability to 'do mathematics' in their chosen language? In 2005, I gave a presentation about a new learning website to the steering group of an organisation that represents the interests of Welsh-medium schools. As part of the presentation, the group was shown how easy it was to toggle between the Welsh and English version of the website. Although impressed with the ease with which this toggling was possible, the group suggested that it would be helpful if such a facility could be disabled by teachers for particular groups of learners. This request came from a very experienced group of bilingual practitioners who could clearly see some disadvantages in affording pupils the opportunity to access an educational resource in what may not be their chosen learning language. In an age where technology is likely to ensure that such opportunities to use two languages for learning become increasingly available, it would appear to be an opportune time to consider under what circumstances, if at all, such flexibility might be desirable.

Bilingual Talk

The previous sections discussed the way in which learning groups are often organised in bilingual schools in Wales and aspects of the 'concurrent' bilingual practices that are often to be found within what I have called 'Option 2' classrooms. The term 'concurrent' is usually associated with situations where teachers systematically translate everything from one language to another for the perceived benefit of at least some learners. In reality, of course, most teachers are far more sophisticated in the way they draw on the linguistic resources available to them and their learners and frequent codeswitching occurs within bilingual classrooms all over the world (Adler, 2001; Banfi & Day, 2004; Heller & Martin-Jones, 2001; Lindholm-Leary, 2001; Martin-Jones, 2000; Mejia, 2004). What happens within these classrooms is not the same in all contexts but is, in each case, shaped by the conditions in which the teacher and learners find themselves and the various external influences upon them. While attempts have been made to describe and categorise the pedagogic strategies that can be adopted for learning and teaching in such classrooms, such as the 'concurrent' approach discussed earlier (Baker & Jones, 1998; Faltis, 1996; Jacobson, 1990; Williams *et al.*, 1996; Williams, 1997), little has been done to examine the communicative practices within such classrooms in Wales.

In a paper published in 2000, I described a study undertaken in 'Option 2' type classes in five bilingual secondary schools in Wales, one of which was School X (Jones, 2000). The study aimed to identify how teachers and pupils drew on their communicative repertoires in mathematics; to provide, through the use of classroom discourse analysis, insights into the limits and potential of bilingual classroom interaction in mathematics and to offer some observations and analytic insights that might encourage teachers to reflect on their own bilingual teaching practices. Adopting a similar kind of approach to that described by Barwell (2004), my intention was to examine the discursive patterns used by the participants rather than trying to understand what the participants necessarily meant or 'had in their head' (Barwell, 2004: 14). The approach drew on Gumperz's notion of the 'contextualisation cue' (Gumperz, 1982: 13). Contextualisation cues can be verbal or non-verbal and indicate a departure from the established or expected pattern of communication. They range from phonological and lexical cues to different types of codeswitching or changes of style. Contextualisation cues can also be at the prosodic (changes of rhythm, stress or intonation) or kinesic (using body language such as facial expressions and gestures). An interactional sociolinguistic approach to classroom discourse looks at the ways in which the participants use various types of contextualisation cues as they negotiate their way through the encounter.

In addition to making audio recordings of some of the lessons, detailed field notes were also made so that key communicative acts such as change of teacher gaze towards particular pupils or the writing of key fragments of text on the board could be noted. Where possible, teachers and pupils were also interviewed either prior to, during or after lessons in order to gain a fuller understanding of the context within which these observation were being made and the communicative resources available to the participants.

This extract below, from School X (discussed in greater detail in Jones, 2000), shows how particular codeswitching practices featured in many of the observed episodes. In the extract below it can be seen that:

- Welsh was used as the main language for managing the class, giving procedural information and so on (see Lines 29, 33 and 35);
- in order to develop the pupils' understanding of the mathematics in question, the teacher repeated explanations and so on in both languages (see Line 16);
- the teacher switched from Welsh to English when referring to calculations or numbers (see Lines 14 and 42–48).

Further, and with reference to the notion of 'contextualisation cues' mentioned above, it can be seen how codeswitching is used to indicate or cue the transition between 'getting the task going' and 'doing the task' (Line 14), to specify a particular addressee (Lines 12 and 38) and to distinguish classroom management utterances from talk about the calculation (Lines 23–35). (For transcription conventions, see Note 2):

```
1        [A WORKSHEET WHICH HAS A NUMBER OF QUESTIONS
         ON IT HAS BEEN SHARED OUT AMONG THE PUPILS. THE
         TEACHER WRITES '25' AND '5' A FEW INCHES APART
         ON THE WHITE BOARD]
5   T:   twenty five per cent .. nawr da ni wedi gwneud hyn o'r blaen
                                  now then we've done this before..
         .. mae dwy ffordd i gael.. mae un ffordd hawdd ok ..
         there are two ways..      one way is easy ok..
```

		meddylia di
10		*you think about it*
		[TEACHER TURNS TOWARDS ONE OF THE PUPILS]
		Owain ?
	P:	[PUPIL RESPONSE IS UNCLEAR]
	T:	**ie gallwch chi wneud huna** ten per cent. ten per cent.
15		*yes you can do that*
		five per cent. **ond mae ffordd haws ok** there's an easier method
		there's an easier way ok
		of finding twenty five percent
	P:	find 50% then halve it
20	T:	ok. ok. fifty per cent then halve it
	P:	**Miss. chwarter e**
		Miss. a quarter of it
	T:	**neu ffendio chawrter e .. nawr te .. cariwch 'malen** .. finish
		or find a quarter of it .. now then .. carry on
25		[A QUIET PERIOD OF 5 MINUTES WHILE PUPILS
		WORK ON THE PROBLEMS ON THE WORKSHEET
		[A PUPIL GESTURES IN A MANNER WHICH
		SUGGEST THAT HE'S FINISHED]
	T:	**llaw lan te .. ti wedi beni ? .. reit o**
30		*hand up then .. you've finished ? .. right o*
		[A NUMBER OF OTHER PUPILS LOOK AS IF THEY'VE
		FINISHED]
	T:	**ti wedi beni ? llaw lawr.. ti wedi beni ? reit .. awn ni**
		you've finished ? hand down .. you've finished ? right ..we'll go
35		**drostyn nhw**
		over them
		[TEACHER TURNS TOWARDS ONE OF THE PUPILS]
		Timothy have you finished ?
	P:	yes
40		[THE TEACHER WRITES '10%' AND '£71.60' A FEW
		INCHES APPART ON THE WHITE BOARD]
	T:	**reit** ten per cent. seventy one pound sixty **reit ..yn iawn..**
		right right.. that's ok
		mae rhai yn ysgrifennu seventy one point six **heb y** nought
45		*some are writing without the*
		. right. the nought must be written ok ..seventy one pounds
		sixty so **mae'r nought...**
		the nought

In a bilingual school, where all speak English but some have only limited proficiency in Welsh, it is understandable that teachers use Welsh for classroom management purposes, since this type of 'incidental' Welsh can contribute to pupils' facility in Welsh. It is also understandable that a teacher might feel it necessary and helpful to repeat a key learning point in both languages. What is less obvious is why a teacher might want to switch to English when talking about numbers or doing calculations. When one appreciates, however, that this kind of codeswitching reflects what is common practice within a community where the majority have themselves received their mathematics education in English, it is easier to understand why some teachers might codeswitch in this manner. School X is located in rural Wales, where, for example, deals are struck at farmers' marts and calculations are made on building sites by native Welsh speakers who were taught to count and calculate in English. Roberts (2000) provides a detailed account of the historical background to such practices.

The codeswitching practices observed at School X are, of course, more than curious features of bilingual talk. These practices are not only key elements of the meaning making process that occurs in bilingual classrooms, they also reflect and shape practices and values within the wider socio-cultural and institutional context. In contrast to the situation in other bilingual settings such as Canada or Catalunya (Heller, 1999; Pujolar, 2000) where such studies have been undertaken, there have only been a handful of ethnographic studies of conversational interaction in classrooms or in other school contexts among secondary school pupils in Wales. These include the study described above and those recently reported by Pike (2004) and Evans and Hughes (2003).

Assessing Bilingual Learners in Mathematics

It is difficult to ensure that summative written tests do not disadvantage bilingual pupils. Over recent years, ongoing teacher assessment or curriculum-related assessment (CRA) has gained widespread international support as a fairer, more reliable way of assessing children from culturally and/or linguistically diverse backgrounds (Cline & Frederickson, 1996). Despite this recognition, written examinations in Wales, as in other parts of the United Kingdom, continue to feature prominently as part of the assessments made for awarding key qualifications at 16+ and 18+ and the proportion of marks awarded under examination conditions seems set to increase over coming years.

The General Certificate of Secondary Education (GCSE) is the main qualification awarded at 16+ in England, Wales and Northern Ireland, and many pupils aim to gain this qualification in up to 10 or more subjects. Since the introduction of GCSEs in 1988, it has been possible for candidates to gain a proportion of the available marks, for any subject, via coursework completed in the classroom or at home, with the remainder of the marks being awarded on the basis of practical or written examination(s). Coursework has provided valuable opportunities for language development as it often involves drafting and refining, as well as offering opportunities for talk about the work. Over recent years, however, concerns about the authenticity of work completed outside examination conditions have resulted in greater emphasis once again being placed on the external, end-of-course examinations and in the case of mathematics, GCSEs will, from 2008, be awarded entirely on the basis of formal, written examinations. Despite the advantages of CRA, it seems that most bilingual learners in Wales will once again be assessed at 16+ solely via formal examinations presented to them in one language. As we return to an era when pupils' attainment will be judged solely on the basis of their performance in such examinations, it seems timely to revisit some of those issues.

Creating matched assessments

There is a growing literature on cross-cultural, cross-linguistic assessment which would appear to be, at least in part, a reflection of the increasing use made of international comparisons of attainment. There are currently around 5000 standardised tests worldwide (Oakland, 2004) and since its formation in 1976, the International Test Commission has worked to develop new standards and guidelines on cross-cultural, cross-linguistic test development and use (Bartram, 2001; International Test Commission, 2000). These guidelines have considerable resonance with what I found from my own investigation into the development of Welsh-medium National Curriculum tests for mathematics (Jones, 1998).

The advantages of ensuring that any cross-cultural assessment is designed collaboratively by representatives of all the linguistic groups that are to take the test have long been acknowledged (Brislin, 1986; Hambleton, 1993). Despite efforts to ensure that words have comparable equivalents in other languages, that contexts are those which all candidates are likely to encounter and that terminology in one language does not 'give the game away' in another, there are many factors that can unexpectedly influence a candidate's response to a question. Some examples of mathematical terminology that are arguably easier to understand in Welsh than in English are *quadrilateral*, which translates to *pedrochr* (four-sides) in Welsh; *plan view*, which translates to *uwcholwg* (above view) and *eighteen*, which translates to *un deg wyth* (one ten eight) or *deunaw* (two nines) in Welsh.

In a paper published in 1993, I described a study where I undertook a comparative analysis of a sample of examination scripts from English-medium and Welsh-medium candidates who had just sat their GCSE mathematics examination (Jones, 1993). An analysis based on responses by 224 candidates to 12 questions was undertaken in order to see whether any specific test items had produced unexpected responses from candidates. Having ensured that the samples were comparable in terms of the overall scores obtained by candidates on the 12 questions, a simple Chi-squared test was undertaken to look for items on which the candidates had performed significantly differently. One such statistically significant item emerged from the analysis: an item that suggested that a less familiar Welsh word was possibly helping Welsh-medium candidates by cueing the correct 'mathematical register'. The question involved the mathematical term *similar*, which translates to the Welsh word *cyflun*. Within a mathematical register the fact that two shapes are *similar* implies that they have the same shape but are of different sizes and candidates needed to use this information to answer the question. Since the word *similar* is also a common everyday word in English, it may not necessarily prompt candidates to consider its mathematical significance. The Welsh word *cyflun*, because it is not a word generally used outside mathematical contexts, is far more likely to do so. Although other explanations for this unexpected item behaviour are also possible, it seems that the word *cyflun* may have prompted more of the Welsh-medium candidates to draw on its mathematical significance to answer the question correctly.

The above study provided an intriguing and rather unexpected example of the way in which 'more difficult' terminology can possibly contribute to a more positive response from candidates in one particular language. Evans (2007) builds on this earlier discussion and provides some further interesting examples of mathematical test item behaviour. He describes, for example, an item that required candidates to identify right angles or *onglau sgwâr* (square angles) and that prompted more correct responses from Welsh language candidates. Like the examples referred to above, *ongl sgwâr* (square angle) probably gives the Welsh-medium candidate a better idea of what to look for. If such subtleties can influence the way in which candidates respond to test items, then it would appear to be essential that all 'high stakes' tests are trialled to identify such rogue items. Over recent years, however, while extensive trialling and analysis have been undertaken on Welsh National Curriculum tests taken by children at 11+ and 14+[3] to identify and eliminate such 'rogue items', no such trialling is currently undertaken on the 'higher stakes' GCSE examinations at 16+ and General Certificate of Education (GCE) Advanced Level at 18+.

Fair testing

The examples described in the previous section serve to highlight the complexity of ensuring that mathematics tests written in two languages provide candidates with the same opportunity to demonstrate their skills and understanding. Beyond test development, however, questions remain to do with the way in which bilingual pupils in Wales can best draw on the communicative resources available to them to demonstrate their true attainment in subjects such as mathematics.

Pupils sitting external examinations in Wales must choose to do so through the medium of either Welsh or English. Candidates are allowed to have both the Welsh and English versions of examination papers on their desks, if their teachers consider this to be helpful to them, but they must answer in either Welsh or English, not both. Anecdotal evidence, however, suggests that candidates, such as those in School X, are not generally encouraged to have both versions of examination papers on their desks since it is argued that having two versions of an examination paper on their desk at the same time could be unhelpful. Since pupils in School X, like many other traditional bilingual schools in Wales, are taught in classes where both Welsh and English are used freely, it would seem appropriate to explore whether there are candidates and occasions when having both versions of an examination paper might be helpful. If some students find it helpful to toggle between Welsh and English on learning websites, it seems reasonable to suggest that some might also gain from having both language versions of an examination paper, or even a bilingual one, available to them.

Concluding Comments

There is a growing corpus of research on bilingual education with many examples of the kind of detailed ethnographic study described in the previous section on bilingual talk. Such studies are providing valuable insights into the ways in which meaning is negotiated and cultural and social values and patterns are transmitted between peers and teachers and across generations. Similarly, there is increasing interest in international comparisons of pupils' attainment in mathematics and in the principles underpinning the construction of valid and reliable cross-cultural assessment tools.

Much of the literature on practice within bilingual mathematics classrooms is concerned with the way in which the use of two languages may facilitate learning (Barwell, 2005b; Clarkson, 2005; Setati & Adler, 2000) and, in particular, the way in which pupils' own language can help them learn through a second language (usually English). As Barwell et al. argue, there are certain contexts in which:

> future research on multilingualism in mathematics education needs to engage with the political role of language. It is only in doing so that our research will make sense to teachers and learners in multilingual classrooms, where language choices and mathematics learning are interwoven and frequently constrained by economic, political and ideological pressures. (Barwell et al., 2007: 119)

Unlike many other examples of bilingual classroom talk considered within this volume and elsewhere, however, Welsh and English have comparable currency within the education system in Wales and the issues considered within this chapter are largely pedagogic. Within Welsh society, Welsh is increasingly recognised as a language that can bring personal, cultural and economic benefits. Indeed, while English may have been seen as 'the language to get on in the world' half a century and more ago, in 21st century Wales, fluency in Welsh is increasingly valued and is, for example, an essential requirement for many key jobs. To consider the challenges bilingual learners and teachers in Wales face as being largely pedagogic may be viewed as being overly simplistic and would certainly not have been the view I would have taken 30 yeas ago. It is clear that a great deal remains to be done in terms of understanding the teaching and learning of mathematics within bilingual contexts across the world. Furthermore, in most cases, teaching and learning must be considered alongside a range of external influences and pressures: economic, social, cultural and political. For those of us working in parts of the world where such external influences are no longer as marked as they once were, the key questions are somewhat simpler and relate to finding pragmatic approaches and solutions to issues, such as managing and enhancing learning, and ensuring that the outcomes of formal assessment procedures accurately reflect the attainment of the bilingual pupils, which they aim to serve.

Notes

1. Welsh-speaking secondary schools as defined in Section 354(b) of the Education Act 1996. A Welsh-speaking secondary school is one where more than a half of foundation subjects, other than English or Welsh and Religious Education are taught wholly or partly in Welsh.
2. Transcription conventions:
   T Teacher
   P(s) Pupil(s)
   Plain font English utterances
   Bold font Welsh utterances
   Italics Translation of Welsh into English
   "Text written on the board
   [ ] [DETAILS FROM FIELDNOTES]
3. National Curriculum tests at 11+ and 14+ have recently been replaced in Wales by teacher assessment.

Chapter 9
Bilingual Latino Students, Writing and Mathematics: A Case Study of Successful Teaching and Learning

KATHRYN B. CHVAL and LENA LICÓN KHISTY

The education of Latinos in the United States is considered to be in crisis (NCES, 2002). This reality has facilitated a disturbing and persistent pattern of underachievement, especially in mathematics. Since a significant number of Latinos are second-language learners (Young, 2002), this is also a crisis of bilingual students. As a result, academically successful, bilingual Latinos are rarely heard of. In this chapter, we present a case study of an atypical fifth-grade classroom (11–12 years old). This classroom is atypical for several reasons. First, although all the students are Latinos, they are successful in learning mathematics. Second, the classroom is in a large, urban, low-income neighbourhood, where the majority of the residents speak Spanish for most business, social and religious activities. In addition, most of the writing in the neighbourhood (e.g. store signs) is written in Spanish, unlike the school environment, where English dominates. By all present social, political and demographic indicators in the United States, children with these characteristics should not be experiencing academic success. This classroom, therefore, challenges broader conventional assumptions about how to educate second-language learners. Later in this chapter, we discuss how the teacher promoted her students' success in mathematics. Of particular interest is the way in which mathematical writing was used for both second-language growth and content learning in mathematics. We begin, however, with brief descriptions of the class, of the teacher's use of writing in teaching mathematics and of one student's development over a school year.

Sara's Classroom

Sara is Latina and has been teaching for over 20 years. She teaches at a school that is nearly 100% Latino, with students of mainly Mexican descent. All her students begin the school year below grade level and leave a year later at or above grade level. This chapter is based on data gathered as part of a year-long ethnographic study of Sara's classroom, designed to understand the effects of Sara's interactions on her students' mathematics learning (Chval, 2001). Over the course of a school year, 119 mathematics lessons were audio recorded. Sixty of those lessons were also observed and carefully documented through field notes. During the study, Sara had a self-contained class of 24 students who represented a wide range of proficiencies in Spanish, English and mathematics.

For programmatic reasons, Sara teaches mathematics in English, but uses Spanish freely to clarify meanings or to motivate her students. Her students have multiple opportunities to help and collaborate with other students and to listen to student explanations in both whole-class and small-group settings. Her students were also expected to produce extensive written work through a process of drafting and revising. Students completed writing assignments in a variety of mathematical genres. For example, students wrote about the procedure for changing an improper fraction to a mixed number, the explanation for the difference between categorical and numerical variables and an argument about the possibility of drawing a right triangle with side lengths of 2, 4, and 6 cm. While conducting the study, it became evident that writing played a significant role in mediating (Vygotsky, 1986) students' learning of mathematics.

For example, Violetta wrote one of the best paragraphs in the class for the first draft of an early assignment of the year, in which students had to explain how to find the missing leg of a right triangle, given the area and the length of one leg (Figure 9.1).

Violetta's writing has several strengths and weaknesses. She used appropriate punctuation and capitalization. She established order by using connectors such as 'first' and 'then'. She explained the purpose for some steps (e.g. 'Then it wil give you the whole rectangle area.'), but not for all of them (e.g. why you draw a congruent triangle to solve the problem). She also included graphics to support her explanation. Finally, Violetta misspelled many words and was not consistent in her spelling errors or use of tenses. It is important to note here that her spelling errors are typical for Spanish-speaking second-language learners (e.g. interchanging b and d to write 'braw' instead of 'draw').

In a writing assignment, seven months after the above writing sample, students were asked to explain how to find the perimeter of a three-quarter circle with a given area of 100 cm^2. This problem is considered to be difficult for this age. Violetta's first draft is shown in Figures 9.2 and 9.3.

Violetta's writing speaks for itself in its overall improvement.[1] Her writing is clear and coherent and reflects a mathematical sense of sequence in her use of 'First, I took ...', 'Next ...', and 'Afther this I divided ...' She offers an explanation for each step and provides in the margin a

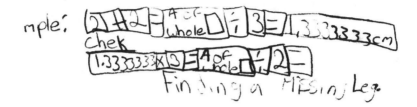

Figure 9.1 Violetta's first draft of the missing leg problem

drawing of the problem shape and refers to it in order to help the reader comprehend what she means. One word, 'build' (built), is worth noting because, again, 'built' has an ending sound ('t') that often is hard for Spanish speakers to hear and remember. Overall, her writing has

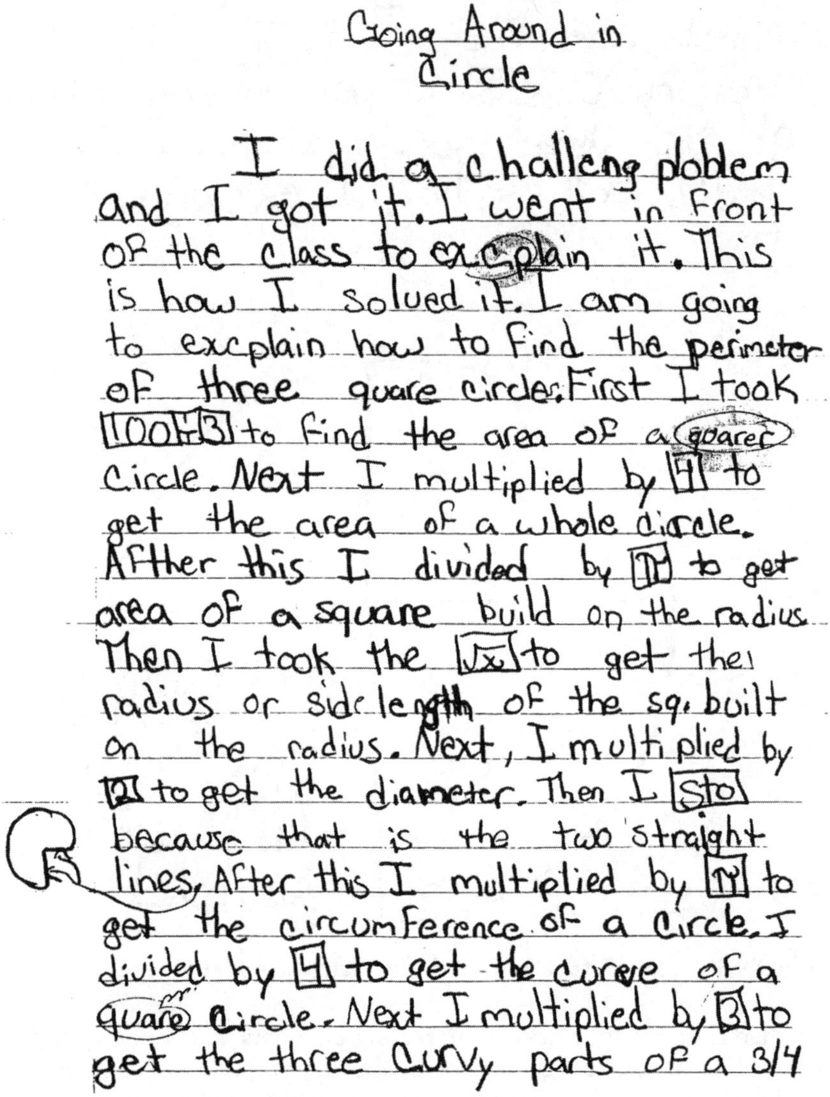

> Going Around in
> Circle
>
> I did a challeng ploblem and I got it. I went in front of the class to excplain it. This is how I solved it. I am going to excplain how to find the perimeter of three quare circle. First I took [100÷3] to find the area of a (quarer) circle. Next I multiplied by [4] to get the area of a whole circle. Afther this I divided by [π] to get area of a square build on the radius. Then I took the [√x̄] to get the radius or side length of the sq. built on the radius. Next, I multiplied by [2] to get the diameter. Then I [stop] because that is the two straight lines. After this I multiplied by [π] to get the circumference of a circle. I divided by [4] to get the curve of a quare circle. Next I multiplied by [3] to get the three curvy parts of a 3/4

Figure 9.2 Violetta's first draft of the semicircle problem (page 1)

improved significantly with only the most difficult second-language acquisition issues remaining.

All students in Sara's class made improvements similar to Violetta's from the beginning of the school year to the end. In the rest of this chapter, we consider how Sara's use of writing enabled her students to learn mathematics and English. We begin by framing our discussion with issues and concepts related to second-language acquisition and to Latinos.

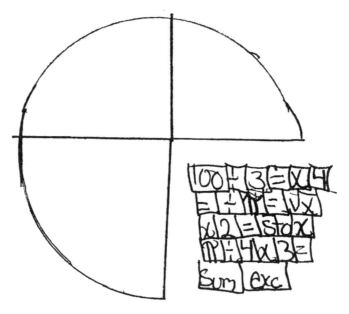

Figure 9.3 Violetta's first draft of the semicircle problem (page 2)

We then describe the various interacting elements that we identified in Sara's classroom that together positively support students' learning. We close with some guiding thoughts about the teaching of content and complex language skills with Latinos and other language minority students.

Framing our Discussion

In the United States, Latinos comprise the largest population of English learners (Garcia, 2001). They may be native-born or recent immigrants, but they come from homes and communities where two languages (Spanish and English) play a prominent role in children's lives. Consequently, mathematics instruction for second-language learners is complex because students must simultaneously learn content and acquire a second language

(English, in our case). However, second-language learners do not need just any level of second-language skill, but rather increasingly more complicated and abstract language for dealing with schoolwork, texts and assessments. Therefore, the development of writing must be an integral part of second-language development and must not be separated from other language development objectives (Peregoy & Boyle, 1993). Research in bilingual education (e.g. Mohan, 1990) also suggests a shift away from less effective instructional approaches that emphasize students simply learning words and other parts of language out of context or as drills. Instead, the development of the second language that is needed for schooling can be better achieved when students use the target language in active and dialogic communication that is purposeful, such as negotiating meanings of texts or solving a problem (Mohan & Slater, 2005).

This perspective of children actively using language does not ignore the importance of the teacher's ability to provide comprehensible input and to serve as a model of the target second language in mathematics (Brenner, 1998; Chval, 2001; Khisty, 1996; Khisty & Chval, 2002). Students *appropriate* language from interacting with their teachers and their peers in collaborative situations and from being guided or assisted in the process. Bakhtin described appropriation as follows:

> The word in language is half someone else's. It becomes 'one's own' only when the speaker populates it with his own intention, his own accent, when he appropriates the word, adapting it to his own semantic and expressive intention. Prior to this moment of appropriation, the word does not exist in a neutral and impersonal language (it is not, after all, out of a dictionary that the speaker gets his words!), but rather it exists in other people's mouths, in other people's contexts, serving other people's intentions: it is from there that one must take the word, and make it one's own. (Bakhtin, 1981: 293–294)

In essence, students only have the potential to appropriate the language needed for academic subjects like mathematics if they have opportunities to hear the language and to use it in communication. Unfortunately, Latino students, like many other linguistic minorities in the United States, too often sit in silence in mathematics classrooms (Brenner, 1998), and thus, do not have opportunities to use and experiment with language, especially the specialized language of mathematics content. Furthermore, their interactions with the teacher typically involve responses to low-level questions that only require simplistic language use and do not facilitate appropriation or significantly advance second-language development (Duran, 1987).

Many cognitive benefits have been attributed to writing in content areas (Lodholz, 1990). For example, writing can help develop concepts or assess understanding. But, in spite of these benefits in mathematics, students are seldom asked to put into words what they did to solve a given problem

(Silver *et al.*, 1990). Moreover, teachers with second-language learners often avoid the use of writing due to students' poor writing skills (Quinn & Wilson, 1997). Consequently, even though research suggests that writing would support the academic and language growth of Latinos, the evidence suggests that they are, in fact, denied access to it.

Writing is considered the most difficult of the four language skills (i.e. listening, speaking, reading and writing) for students to acquire proficiency. Writing requires extraordinarily deliberate analytical thought and linguistic action on the part of the child (Vygotsky, 1986). As a result, writing is the last of the four language skills to develop among second-language learners who are writing in their weaker academic language (Peregoy & Boyle, 1993). To become proficient in writing, especially in a second language, requires students to have frequent opportunities to write and receive constructive feedback from the teacher; this not only improves writing skills, but also promotes second-language acquisition (Peregoy & Boyle, 1993). However, writing in content areas such as mathematics is made more difficult for second-language learners because mathematics includes different genres or types of writing seldom used in other content areas, such as procedure (i.e. description of methods) and exposition (i.e. arguments) (Mousley & Marks, 1991). While there is little agreement as to what specifically makes up mathematical writing (Morgan, 1998), we can recognize it by characteristics such as the special syntax (e.g. mathematical word problems; see Barwell, this volume), syntactical forms such as 'if ... then' statements, use of symbols and an emphasis on conciseness (Monaghan, this volume; Morgan, 1998; Pimm, 1987). Also, language use in mathematics goes beyond mere words or terminology. Mathematics involves ways of thinking and valuing as well as ways of expressing mathematical meanings (e.g. use of examples, conciseness and multiple representations). These language characteristics (i.e. genre, specialized language of mathematics) comprise the cultural tools for expressing mathematical ideas and students have to make choices about how they use these tools. An effective writer of mathematics needs to be able to select the appropriate genre (e.g. procedure, exposition, etc.) and use it with appropriate language including the specialized language of mathematics. For mathematics instruction, this means that writing should genuinely be a communication tool connected to supporting dialogue and meaningful activities (Mohan & Slater, 2005), so that understandings can be appropriated (Bakhtin, 1981). In the next section, we demonstrate and explain how this occurred in Sara's classroom.

Writing in Sara's Classroom

To fully understand students' achievement in Sara's classroom, we had to consider the writing process: how was writing mathematically in a second language (English) developed and how was writing mathematically connected to the learning and understanding of mathematics. To address these questions, we did a close analysis of Sara's instructional moves and interactions, as well as of students' writing. Through qualitative analysis methods (Glaser & Strauss, 1967), we identified three significant patterns:

(1) Sara created a culture that valued writing in mathematics;
(2) Sara immersed her students in an environment filled with rich mathematical language and interactions;
(3) Sara created a dialogic relationship through written questions and other feedback to facilitate mathematics and language learning.

Pattern 1: Sara created a culture that values writing in mathematics

One of the key patterns in Sara's instruction is the tremendous level of thought and effort she expended in creating a culture that values writing. The most striking evidence of this is that she had students begin writing in mathematics from the first day of school. Sara's instruction also made explicit to students that she herself values writing. Through statements like 'I notice that Javier is writing everything he knows about the figure that is in front of him and that is very important, a very important thinking tool' (Day 11), she set the class cultural norm that writing is important. She modelled good writing as she wrote sentences on the chalkboard and as she wrote comments on student assignments; she devoted time to writing in her mathematics classes, in her preparation of lessons and in the assessment of student writing; and she repeatedly discussed the importance of writing with her students. For instance, in the first 12 lessons of the school year, Sara prominently, and in context, used the words, 'tell', 'dictate', 'explain' and 'write', to encourage students to 'talk' during the whole-class discussions.

The following list briefly gives an overview of additional specific practices that demonstrate how Sara created a culture around writing:

- Sara provided time in class for students to write on a daily basis.
- Sara required as many as six drafts to meet her expectations for a final product. In our conversations with Sara, she stated, 'The early drafts permit the child's voice to emerge – who they are and what they are trying to say. Drafts 4 and 5 begin to focus on conventions'
- Sara made mathematical writing a public process. She used multiple techniques to engage students in the writing process. These techniques included reading samples to the class, facilitating a brainstorming discussion of the appropriate words and ideas for specific writing assignments, asking students to read their own work out loud, requiring students to read and comment on each others' drafts,

facilitating whole-class editing of sentences or ideas in a specific paper and asking students to brainstorm writing ideas with their peers.
- Sara created writing assignments that genuinely asked students to communicate about what they were learning in mathematics. The assignments were not simply an extra activity for students but became a natural part of what it meant to do mathematics.
- Sara used a clear five-category writing rubric to grade each final product. Students received a score for 'focus', 'support', 'organization', 'conventions' and 'overall'. She incorporated this strategy from language arts.
- Sara wrote questions on student assignments leading students to be more explicit about their ideas. This practice is described in greater detail below.

Pattern 2: Sara immersed her students in an environment filled with rich mathematical language and interactions

Sara's classroom was filled with active academic conversations (both written and verbal) between students and teacher, where students encountered new words and expressions, negotiated meanings, had opportunities to speak and write them and thus gained control of them or appropriated them. Sara guided and supported her students in this process by (1) speaking and writing sophisticated words and expressions; (2) setting a purpose for writing and (3) building meaning for these words in context. Sara did not reduce the curriculum's level of complexity, including its language (Moll, 1990). Instead, she created an environment in which students experienced and used rich words and language in context. Students not only heard this language, but they saw it written on the chalkboard (simultaneously, as Sara said it), on student papers and in written feedback on their assignments. For example, during the first two weeks of school, Sara assigned students to write an explanation for solving a mathematics problem. Sara wrote comments on each successive draft of the assignment. The following examples of Sara's feedback written on student assignments demonstrate how she immersed students in an environment filled with sophisticated talk, right from the beginning.

> *Verify* your results.
> *Combine* the areas.
> What does that number *represent*?
> Your work is *deteriorating*.
> Use your *power of observation*.
> *Clarify* this example.
> Where do you *build* a *congruent* triangle?
> Next step, check *conventions*.

Some of the words Sara used are cognates for words in Spanish; for example, 'combine', 'represent' and 'clarify' are cognates for the Spanish words *combinar*, *representar* and *clarificar*, respectively. Students, therefore, can use their knowledge of their home language to decipher new meanings.

Sara also set a purpose for writing. For example, over the course of the first 12 mathematics lessons of the school year, students discussed and learned about the area and perimeter of right triangles. Students struggled with the problem: 'A right triangle has an area of 500 sq. cm and a leg of 200 cm. Calculate the length of the other leg. Draw a sketch of the triangle'. To help students clarify their thinking for this problem, Sara asked her students to write an explanation for how to find its solution as follows:

> How do you explain to someone how to find the missing leg of a right triangle, when you know the area and the other leg? Explain how you did all that. I want to take this home and I want to show it to one of my friends who doesn't know how to find the area of a right triangle let alone a missing leg.... But I want her to know how to do it and I want her to read your papers and *if she understands what you wrote, then you've done a good job explaining* (italics denotes emphasis in her voice). [Day 9]

Sara did not simply ask students to write a report of what was done to solve the problem. Rather, she asked students to *communicate* so clearly that anyone – especially someone who is inexperienced with the mathematics – could understand it. Also, to assist the students with their first drafts, Sara asked the students to identify words and statements that could be included in the written explanation. Sara then wrote the resulting list (e.g. congruent, build, leg, other, area, whole, divide, multiply, add, think, for example and sketch) on the chalkboard.

Fifteen of Sara's students completed four to six drafts of the 'missing leg' writing assignment over a three-week period. In analyzing and comparing the first and fourth drafts of this assignment, we found considerable growth in the use of four terms (i.e. original, rectangle, right triangle and whole) as presented in Table 9.1.

More importantly, Table 9.1 shows in the first column how often Sara used each of these words during whole group discussions of the first 12 lessons. This frequency is noteworthy because it shows how she exposed her students to the language of mathematics and provided a supportive environment where students could appropriate the language as their own. Of course, some words were spoken more frequently than others because they were needed more. For example, the students were solving area problems everyday so that the word 'area' was used most frequently within this context. On the other hand, the word, 'original', was used a small number of times, specifically in the case of solving the 'missing leg' problem described

Table 9.1 Frequency of word use by Sara and her students

Word	Frequency of Sara's use of words during the first 12 lessons	Number of students who used word on first draft	Number of students who used word on final draft
Original	27	0	8
Rectangle	395	6	12
Right triangle	78	4	13
Whole	215	7	11
Area	699	13	15
Congruent	143	14	15
Divide	206	14	15
Leg	442	15	15
Missing	101	13	15
Build	31	8	8
Add	69	14	12
Example	23	9	10
Multiply	90	8	10

above. This word was helpful for distinguishing the first, 'original' triangle from the second, congruent triangle needed to create a rectangle.

Sara did not just repeat the above words frequently in class discussions; instead, she used the context to build meaning for each of them. For example, 'congruent' was a word that was 'foreign' to Sara's students when they entered her classroom. Based on classroom observations, the students had trouble pronouncing it, they resisted speaking it and they did not seem to know its meaning. The following excerpt from a class discussion illustrates how Sara used the context to guide students in developing meaning for this word. In this excerpt, 'S' refers to a response from one student and 'chorus' refers to a response from several students in unison. It should be noted that this episode occurred during the first week of school and, therefore, the discourse pattern reveals typical limited responses from the students; they had not yet developed proficiencies or habits for extended speaking. It should also be noted that this was *not* the first time 'congruent' was discussed. The underlined words are our emphases:

1 **Sara:** What do I need to do to the 24, to get the area of that right triangle?
2 **S:** Divide by two.

3	**Sara:**	Why do I divide it by two?
4	**S:**	You have two triangles.
5	**Sara:**	I have two <u>congruent</u> triangles here. Two <u>equal</u> parts, two <u>exact</u> triangles. I want only the area of my original triangle, ACB. Then I'm going to divide this by two. And what will my answer be?

...

9	**Sara:**	Number three. Would you please read that, Julia?
10	**Julia:**	The triangle and its ...
11	**Sara:**	Congruent.
12	**Julia:**	Congruent [struggling] ...
13	**Sara:**	Look at that word everyone. Congruent. <u>What does that mean</u>?
14	**Javier:**	Like another copy.
15	**Sara:**	An <u>exact copy</u>. Because here, look here is the circle. Is this circle congruent to that circle?
16	**Chorus:**	No.
17	**Sara:**	No, they're not <u>exact copies</u>. They're similar, they're both circles, but they're not <u>exact copies</u>.
18	**Chorus:**	Yes.
19	**Sara:**	How about this one and this one?
20	**Chorus:**	Yes.
21	**Sara:**	They appear to be congruent to each other. I agree. They appear to be congruent. But this one and this one are not congruent, are they?
22	**Chorus:**	No.
23	**Sara:**	So, congruent means an <u>exact copy</u>. Javier, you are super right.

Notice that Sara repeated the student's answer from Line 4, but added 'congruent' (Line 5). She began by using 'equal' and 'exact' to describe congruent (Line 5), but Javier provided a different description, 'another copy', for congruent (Line 14). Sara refined Javier's answer by combining her description with Javier's description to create 'exact copy' (Lines 15, 17 and 23), but she gave Javier all the credit (Line 23).

From this point forward, Sara consistently referred to 'congruent' as a 'copy'. Over the course of the next several lessons, Sara put these two words together, 'congruent copy'. As the students became comfortable using the word 'congruent', she removed the word 'copy' and began to use more precise language such as 'congruent triangle'. As a result, we see how Sara guided her students' appropriation of this word by using the word frequently in the context of solving problems and by creating meaning for it. The evidence of this appropriation is the students' use of the word in their writing. Eventually, 'congruent' was a word that appeared in the writing of every student in the classroom. However, we want to point out that the issue of language development involves more than just acquiring words. It concerns the expression of mathematical ideas. Words are a small part of the meaning-making discourse process, as we discuss in the next section.

Pattern 3: Sara used written questions and other feedback to facilitate mathematics and language learning

When Sara gave a writing assignment, her primary goal was that students clearly communicated the mathematical concept or process. Initially, she did not mark spelling or grammatical errors. Instead, she began a conversation by posing many questions in response to the writing, as seen in Javier's second draft in Figure 9.4.

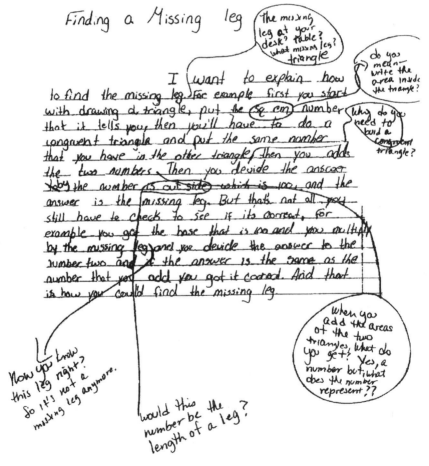

Figure 9.4 Javier's second draft

Sara's questions to Javier, taken from this example, include:

Javier (Draft #2)

> The missing leg at your desk? Table? What missing leg?
> Do you mean—write the area inside the triangle?
> Why do you need to build a congruent triangle?
> Would this number be the length of a leg?
> When you add the areas of the two triangles, what do you get? Yes, a number but, what does the number represent?
> Now you know this leg right? So it's not a missing leg anymore.

Sara asked questions intended to guide students' clarification of their thinking, such as 'What missing leg?' However, this clarifying question also initiated a conversation between the student and teacher (e.g. Javier responded to Sara's question by writing, 'triangle'). This reciprocal nature characterizes a written conversation. Sara even anticipated one of Javier's responses. She asked, 'When you add the areas of the two triangles, what do you get?' She recognized that Javier already had written the word 'number' six times in the draft and she anticipated he would reply, 'a number'. Sara communicated to Javier the need to extend his thinking to the meaning behind that number as she continued, 'But what does the number represent?' Through such questions, Sara started a written conversation with Javier, and used this conversation to guide Javier's writing, as well as his mathematical thinking. Through this conversation, she brought Javier into the cultural practice of mathematics and into writing to communicate.

Sara also assessed students' needs through writing and adjusted her guidance accordingly. For example, Sara initiated conversations with Matthew on his first and second drafts. However, on Matthew's third draft (Figure 9.5), he did not address Sara's questions and he omitted the example that was given on his second draft.

As a result, Sara provided general advice as follows.

> #1 Not clear because many details are missing.
> #2 You did not include any examples so I cannot see what you mean.

You would have a better explanation if you:

1) reread your work; 2) add details; 3) draw a sketch; 4) write keystrokes[2]; (5) think a little more.

It is interesting that some of these comments (#1, #2 and '3' in the last section) also refer to what is considered good mathematical practice. As the students began to more clearly communicate their mathematical ideas, Sara's guidance turned to the improvement of their writing skills. For example, on Alejandro's fourth draft, Sara suggested that he add some

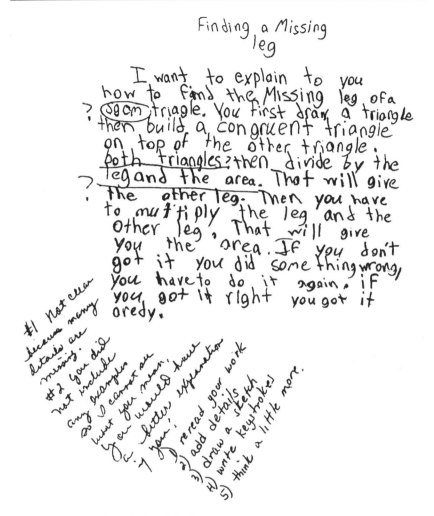

Figure 9.5 Matthew's third draft

information and she made the decision that he was ready to focus on the conventions of writing. She wrote:

Alejandro (Draft #4)

> Outstanding. You did it in only four drafts. I think just about anyone will understand how to find a missing leg!
> Excellent. It is perfectly clear. Next step—check *Conventions*. Conventions? Yes. This means, spelling, punctuation and capitalization.

It is clear Sara expected Alejandro to find his spelling errors, and he had several resources with which to do this, including his peers, the dictionary and the 'spell check' feature on the computer. On this assignment, Alejandro misspelled 'combind' and 'explane' – typical mistakes of English language learners. Based on comments Sara made during informal interviews as well as directly to the students, Sara was not concerned that this particular assignment be a perfect document. She was interested in developing the children's understanding of mathematics and their ability to think and write about their ideas. However, she also recognized that students must also strengthen their English skills in using conventions.

Closing Thoughts

Sara's students significantly improved their mathematical writing and their use of the academic language of mathematics over the course of the school year. Moreover, her students made tremendous gains on standardized mathematics tests, demonstrating their learning of mathematical concepts and problem solving. For example, Table 9.2 compares the median grade equivalent on the Iowa Test of Basic Skills (ITBS) Mathematics Total for Sara's students with other fifth grade students in their school, district and nation.

As shown in the fourth grade column, Sara's students performed below their peers in their school, district and nation before they entered Sara's classroom. In fact, only five of the 24 (20.8%) performed at the 4.8 level or above. After just eight months in Sarah's class (fifth grade column), her students outperformed the other three groups and 15 of the 24 (62.5%) performed at the 5.8 level or above.

Sara's case demonstrates the results of using mathematical writing within a model of integrated content learning and language development (Mohan, 1990; Mohan & Slater, 2005). She provides a critical example of how to *support* students as they simultaneously do advanced mathematical work and acquire a second language. She maximizes contextual supports such as drawings, calculators and other representations that complement other supports such as teaching for meaning, using students' home

Table 9.2 Comparison of median grade equivalent on the Iowa Test of Basic Skills (ITBS) mathematics total for Sara's students with other fifth grade students in their school, district and nation

Comparison groups	*End of fourth grade*	*End of fifth grade*	*Gain*
Sara's class	4.3	6.1	1.8
Other fifth graders in Sara's school	4.6	5.8	1.2
District	4.6	5.6	1.0
National norm	4.8	5.8	1.0

language and thinking as learning resources, conversing through writing and modelling through her own talk.

Specifically, after analyzing Sara's instructional moves and interactions, we found three significant patterns with Sara's instruction: (1) Sara created a culture that valued writing in mathematics; (2) Sara immersed her students in an environment filled with rich mathematical language and interactions; and (3) Sara created a dialogic relationship through written questions and other feedback to facilitate mathematics and language learning. These three patterns were interdependent, meaning all three were critical for success: removing one would influence the other two. For example, if Sara did not value writing in mathematics, she would not have invested significant class time for mathematical writing or personal time to provide written feedback on students' writing. If Sara did not speak and write sophisticated words and provide opportunities for students to speak, write and negotiate their meanings, then students would not have appropriated them.

For Sara, learning is accomplished in part by creating a *culture* where students are expected to communicate mathematically. For Sara, writing is foremost a tool to purposefully accomplish that communication (Khisty, 2001; Moschkovich, 1999b). Writing, then, is one form of social activity and dialogue needed to learn (Vygotsky, 1986). For Sara, writing is an integral part of second-language development and does not stand apart from the overall learning of mathematics. Throughout the year, there was virtually no evidence of instruction that focused on practicing writing skills or memorizing vocabulary, where words were disconnected from their function for communication and meaning making. Moreover, Sara did not reduce the curriculum's level of complexity, including its language (Moll, 1990). Instead, she created an environment in which students experienced and used rich words and language in context. She facilitated active academic conversations (both written and verbal), where students had opportunities to speak, write and negotiate meanings in collaborative situations. Written and verbal conversations about specific writing assignments took place over several weeks, giving the students sufficient time to interact with Sara, respond to her feedback and gain control or appropriate mathematical language and practices.

The characteristics of Sara's instruction cannot be reduced to single strategies; they must be taken together, integrated, and applied in multilingual mathematics classrooms, if the crisis of Latino students' education is to be halted. Fortunately, Sara teaches us that it is possible.

Notes

1. You may notice Violetta's unique use of calculator keystrokes within her explanation. We have chosen to narrowly focus our discussion on general aspects of mathematical writing. However, for further discussion see Chval (2001).
2. If students used calculators to solve a problem, they were required to include a record of the calculator keystrokes that were used.

Chapter 10

Mathematics Teaching in Australian Multilingual Classrooms: Developing an Approach to the Use of Classroom Languages

PHILIP C. CLARKSON

In Australia, immigrant students who are English language learners (ELLs) need to learn the culture of Australian schooling (Clarkson, 2006a). They also need to learn to communicate quickly and effectively in English, since this is the dominant teaching language. This is an important issue in many Australian classrooms in the large cities, where there can be students from several different language backgrounds in one class. For example, it would not be surprising in many urban Australian classrooms to find that one student is from an immigrant family that recently came from Vietnam. The first language she learnt at home was Vietnamese and this is the language normally used in her home. She only started to learn English when she attended school. Now in Year 2 (7–8 years old), she is able to have conversations with other students. She uses her first language with her Vietnamese classmates both in and outside of the classroom as a matter of choice, as well as in Saturday school held in a local church hall. At the same table is a boy whose family came from Greece 20 years ago, and although he was born in Australia, Greek is still spoken at home and with most family friends. Most family activities occur within a close-knit Greek community. Another boy on this table comes from Somalia. He is 10 years old (old for this year group), but only started school 12 months ago when his family finally entered Australia as refugees. And so on around the room where we may well find something like six or more languages represented, with students having various levels of proficiency in their languages.

The various State Education Departments in Australia have provided help for ELL students for many years, mainly directing this help to the learning of English as a *communicative* language, that is, as a language of everyday conversation. This is to the benefit of such students and often, within two years or a little more, they gain good proficiency in communicative English. However, the need goes beyond day-to-day communicative English. In the past it was thought that once students mastered communicative English, they would have sufficient proficiency in English for learning different subjects like mathematics. Clearly, proficiency in communicative English is important in a learning context. But there is also a need for these students to be able to communicate in *academic* English, that is, the specialised form of English used in subjects like mathematics. This process can take several years longer than learning communicative English (Cummins, 2000a). To be only satisfied with ELL students gaining proficiency in communicative English in a reasonably short time is to undersell the potential of these students. There is a need for mathematical learning situations in which ELL students also gain proficiency in the academic English used in mathematics.

The Role of Language in Learning Mathematics

The role of language in mathematics learning has gradually become a feature of mathematics education research over the last three decades (Austin & Howson, 1979; Ellerton & Clarkson, 1996; Ellerton *et al.*, 2000; Halliday, 1978; Pimm, 1987). The link between language and mathematics learning is now seen as a crucial issue in mathematics teaching. Within this research area there sits the more specific issue of students' learning when the student can converse in two or more languages.

To perform well at school, students have to master the teaching language, which, in Australia, is normally English, the dominant societal language. However, it has been shown that multilingual students outperform others academically if they are proficient not only in the teaching language, but also in their first language (e.g. Cummins, 1991, 2000a, 2001). Over the last 25 years, this finding has been shown to apply to bilingual students doing mathematics (Clarkson, 1992, 1995; Clarkson & Dawe, 1997; Clarkson & Galbraith, 1992; Cocking & Mestre, 1988; Moschkovich, 2002; Secada, 1992). Continuing work also shows some of the reasons why bilingual students switch languages when doing mathematics (Clarkson, 2006b; Clarkson & Indris, 2006; Parvanehnezhad & Clarkson, 2008). It is one thing to understand something of the impact of language in general on the learning of mathematics and, in particular, the crucial role that language plays for multilingual students learning mathematics. However, the implications that these results have for teaching have not been explored sufficiently. It is with the teaching of mathematics to ELL students that the remainder of this chapter is concerned.

The lack of research on teaching in multilingual classrooms particularly applies to teaching mathematics. Of hundreds of articles published in four leading journals in mathematics education[1] from 2000 to mid-2006, there were less than 10 that focused on the teacher's role in such situations. However, the following four notions from these studies are worth noting:

- Teachers should encourage the use of different types of language, such as informal talk in students' first language leading to more formal mathematical talk in the language of teaching.
- Tracing the language paths of students in such complex multilingual situations is critical.
- Informal or exploratory talk often occurs in students' first languages. This can often lead to 'broken communication' when the teacher does not share the students' first language. Hence helping students to move to more formal mathematical talking and writing, which often involves a switch to the language of the classroom, can be fraught with unknown linguistic challenges.
- Teachers need to use academic mathematical language in verbal discourse and promote an expectation that students will come to use such language. Findings suggest the students do, in the end, use formal mathematical language if they see the teacher using it consistently.

In the next section, a model of the use of language in mathematics classrooms is discussed. The model is described for the monolingual teaching context, before being extended to include multilingual contexts.

Language Aspects of Mathematics Teaching Strategies

The recognition of the influence of language use on mathematics learning gave rise some time ago to a model suggesting how teachers could encourage students to use language effectively (e.g. Clements & Del Campo, 1987; National Council of Teachers of Mathematics, 1989). This model suggests that students progress through different types of language (Figure 10.1): from informal language, through more mathematical structured language to academic mathematical language. Informal language includes frequent use of idioms quite specific to the particular age group, and perhaps specific to a city or a school, as well as shorthand, abbreviated language. More mathematical structured language is that which teachers often use in school. In informal conversation, it is common for speakers not to use full sentences. In more structured language there is a rise in speakers using full sentences. When using the written language, the use of informal jottings decreases as students move into a more structured type of writing. Finally, academic mathematical language consists of the more specialised language used in the field of mathematics that has been developed to allow a sharing of precise mathematical thinking (Pimm, 1987).

In this model, the informal language of students and the corresponding contexts with which they are familiar and from which their language

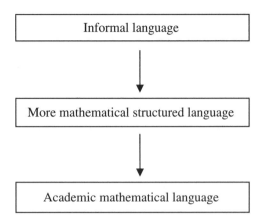

Figure 10.1 Language use in mathematics learning

draws its meanings are used by the teacher to begin a discussion of a concept, procedure or issue. Such a language type might not be capable of exploring extensively the nuances of the mathematics under discussion. As the discussion progresses, students are encouraged by the teacher to move from the informal to more mathematically structured language and finally to the precision of academic mathematical language. Hence, students will be guided to be careful how they use such words as 'half' (not the everyday usage 'part of a cake' and certainly not 'taking the biggest half'), as well as being encouraged to use specialist mathematical words such as 'ellipse', 'parallel', and so on.

The model is reflected in some Australian curricula. In the Queensland curriculum, for example, terms such as 'everyday language', 'familiar mathematical language' and 'basic mathematical language' are used in the guidelines, which indicate the level to be reached by the end of Year 3 (8–9 years old).[2] For the guidelines for the end of Year 5 (10–11 years old), only 'basic mathematical language' is used, and for the ends of Year 7 (12–13 years old) and Year 9 (14–15 years old), the only term used is 'mathematical language, symbols and conventions'. In general, far more use is made of terms such as 'language precision', 'difference to everyday language' and 'symbolic context of language' in such documents. Hence, although there is some recognition of the suggested flow of language use from informal to academic, the emphasis is more on progressing speedily to the academic language of mathematics (e.g. Queensland Studies Authority, 2004; Victorian Curriculum and Assessment Authority, 2002).

Although the model provides useful guidance for teachers, it is limited. It is surprising that the model has not been elaborated to reflect the far more dynamic manner in which both teacher and students use language. One way to capture this dynamism is to use double-headed arrows (Figure 10.2). Knowing how conversations in classrooms develop, there

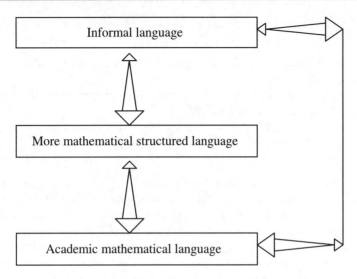

Figure 10.2 A modification to the language use in mathematics learning model

seems to be a strong possibility that at times the intermediate category will be omitted. Including a double-headed arrow that links the top and bottom cells in Figure 10.2 overcomes this problem. The use of double-headed arrows suggests that both teacher and students may well move backwards and forwards between the language types, although the predominate movement of language use over time is in a downward direction, resulting in the conversation ending with the use of more formal mathematical language. The varying size of the arrows is intended to show this dominant flow. The time span of the 'the conversation' may span several lessons as the class comes to a deeper understanding of a particular topic. The object of this model is that students will be able to converse at a number of levels regarding mathematics, and when called for, be able to use, with meaning, the academic mathematical language embedded in the subject.

Although there has been wide acceptance of an approach to mathematics teaching that emphasises the importance of language, it seems that the model was developed assuming that all classrooms are monolingual sites of learning. This model is now considered again, this time extended to assume a multilingual site of learning.

The Beginning of a Teaching Model for Multilingual Contexts

The notion of the multilingual context of many classrooms can itself be misleading. There is no such a thing as 'a' context. There are many different contexts. Some of these contexts, elaborated elsewhere in more detail (Clarkson, 2004a), are

- monolingual teachers teaching a mixture of monolingual and multilingual students;
- monolingual teachers teaching classes of multilingual students all speaking the same languages;
- multilingual teachers teaching multilingual students in a language different to any of their first languages and
- multilingual teachers and multilingual students who share a language.

Clearly teaching mathematics in multilingual contexts is not straightforward. Teachers need to cope in situations where they will not have full management of the discourse, unless they too are proficient in the students' language(s) as well as the teaching language. However, the flow from exploratory verbalising of ideas to formalising them in a rich mathematical language seems to be given across the contexts. How to manage the language flow is the issue that needs further research. A modification of the model developed for monolingual teaching contexts (Figures 10.1 and 10.2) is needed to take account of the multilingual contexts of mathematics classrooms. One approach, developed over a long period by the author in Australia, is outlined in Figure 10.3.

It is worth considering whether some cells in the model shown in Figure 10.3 are non-viable or of limited usefulness for teaching mathematics in any context or in particular mathematical contexts. For example, if

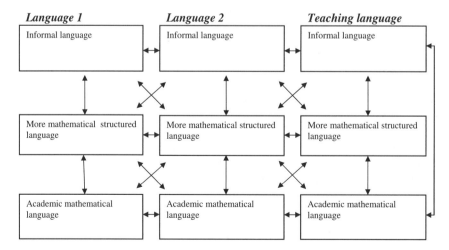

Figure 10.3 A model for language use in mathematics learning for multilingual students, with the overall flow of language downwards

Language 1 is a village language in Papua New Guinea, it is probable that the bottom-left-hand cell will be empty. For example when teaching in early primary years in the Whagi Valley of the Western Highlands, teachers are encouraged to use English, *Tok Pisin* (the national lingua franca) and *Whagi*, the local language. In *Whagi* the notions of 'whole', 'half' and 'quarter' can easily be expressed. But it is very difficult, if not impossible, to talk with meaning about other fractional quantities (Charly Muke, personal communication). However, this cell will not be empty for students in Australia who speak Vietnamese. In Vietnamese, all the mathematical notions that occur in the primary years in Australian schools can be expressed with clarity. Hence consideration of the viability of cells will be dependent on the language.

Another interesting issue is that of which of the arrows or linkages the teacher should encourage. Some potential linkages are not shown at all in Figure 10.3, such as a linkage between the top-left-hand cell and the bottom-right-hand cell. One suspects that there would be little call for this linkage. It may be more useful for teachers in specific language contexts to explore particular linkages rather than others, even though students may use some linkages occasionally. Halai (this volume), for example, highlights the difficulty faced by teachers in navigating what language linkages they should encourage. Working in Pakistan with a teacher and students who spoke Urdu as their first language and English as the official teaching language, students switched between their languages at the informal stage (to move from the top-left-hand cell to the top-right-hand cell). Students were, however, confused by some of the informal everyday English the teacher used (located in the top-right-hand cell). As the teacher encouraged students to move down the right-hand column of Figure 10.3, the students' further understanding of the mathematical problem was compromised by this initial confusion. Hence, the general model needs to be adapted to each context.

In mathematics education research, the importance of recognising specific multilingual contexts has only begun to be recognised (Barwell, 2005e; Clarkson, 2004b). Without recognising the plurality of this issue, there is a danger that a model for teaching that may be useful in one such context can be assumed to be applicable in all multilingual contexts. This is the same mistaken assumption made with the translation of teaching strategies from monolingual contexts to multilingual ones. For example, Setati and Adler (2000) developed a model situated in the context of South African schools. Although there are many points of similarity between their model and the present formulation, different contexts of the two teaching situations are also important. In Australia, we have not had to deal with the aftermath of an Apartheid era; nor do we have many teachers who are multilingual. It is interesting, therefore, that the flow of language use is similar in the two models. While differing contexts may produce similar models, however, the reasons for their use and the nuances that they bring out may well be different.

In Australian Urban Classrooms

Although the importance of language is now recognised in the teaching and learning of mathematics in Australia, most proposed teaching strategies assume a monolingual context in the classroom. There are few suggestions of how to modify one's teaching for multilingual contexts. Few teachers, for example, recognise that students may use their home languages when they solve problems.

In mathematics classrooms such as the one described at the start of this chapter, students do indeed use their non-English languages to do mathematics. Clarkson (1996) briefly examined the knowledge of mathematical language that bilingual Vietnamese and Italian elementary school students in Melbourne and Sydney had in their home language. For most of these students this knowledge was fragmented. For these students, a dotted line joining the bottom cells in Figure 10.3 would be a better representation. Similar findings have been found with Iranian students in Australia (Parvanehnezhad & Clarkson, 2008). Even so, when faced with solving mathematical problems, such students at times did use their home language for a variety of reasons. As one Vietnamese girl (S1) aged 4 years suggested during an interview with the author (I):

I: Did you reread the whole problem, or did you just say 4 + 6 straight away?
S1: I just said 4 + 6
I: Straightaway. Did you do that in English or Vietnamese?
S2: Vietnamese
I: Do you do all your numbers in Vietnamese?
S2: Yes
I: And why do you choose Vietnamese?
S2: It was sort of like easier.

Or as one of the boys (S2) said:

S2: My mum and dad told me a Vietnamese way so I did it the Vietnamese way.

It is hard for teachers to adapt to teaching in multilingual contexts like the one described above if that context is over-simplified in the curriculum documents they use to guide their teaching. This is the situation for many Australian teachers when teaching mathematics in multilingual contexts. As part of every Australian curriculum, there is a section on the teaching of English as a second language (ESL). Thus there is clear recognition of the multilingual context of many classrooms. But an inspection of the ESL sections did not reveal any references to mathematics. In turn, the mathematics curriculum documents make no references to ESL or bilingual learners. This suggests that there is no official recognition of the multilingual contexts of classrooms when mathematics is taught.

Towards the Implementation of Such a Model in Australian Urban Classrooms

Even though there is little recognition in Australian curricula of the importance of language for mathematics learning and no recognition of the impact of multilingual contexts, there are a number of strategies that together may resolve into a useful model to move this situation forward for teachers in Australian urban classrooms. Four possible classroom strategies will be outlined. The first suggests that teachers need to have a clear idea of the languages that are represented in their classroom and hence a mapping exercise is called for. The second refers to Figure 10.3 as a way to track students' uses of their languages and suggests that teachers start to engage in encouraging students to explicitly explore the use of different types of English and their first language. The third looks to a key potential resource, engaging the expertise of ESL colleagues. The fourth focuses on engaging the students' home communities as a resource for their mathematics learning. In this way, teachers can engage with the source of their students' first languages and their idiomatic use of languages (the first row of Figure 10.3).

Engaging with the Language Complexity of Classrooms

Perhaps a first key strategy for teachers teaching in a multilingual context is for them to map the languages that are represented in their classroom, and record each student's competencies in the languages they use. To start charting the language complexity of a classroom it is useful to consider the interacting sources of language. In general, one of these is the students' language or languages. Then there is the teacher's language or languages, as well as the official teaching language or, less often, languages. In typical Australian classrooms, such as the one I have described, this complexity is reduced to some extent, since the official teaching language is only English and, in most cases, the teacher is monolingual in English.

Teachers also need to consider the 'types' of language referred to in the three rows of Figure 10.3. The 'informal' language that students use in informal situations, such as when playing together outside of the classroom, and in the classroom between students, are often known to teachers as 'student chatter' and exist in any language. 'More mathematical structured' language will also exist in all languages, at least to some degree. However, 'academic mathematical' language may exist in

Table 10.1 Mapping languages used in Australian urban classrooms

	Language of instruction: English			Student's first language			Student's second language		
	A	B	C	A	B	C	A	B	C
	Informal language	*More mathematical structured language*	*Academic mathematical language*						
Teacher: L1:...... L2:......									
Student 1 L1:...... L2:......									
Student 2 L1:...... L2:......									
...									

students' home languages, but for some languages, at least for some areas of mathematics that may be studied in Australian classrooms, it may not be well developed.

The first attempt to show these components of the teaching situation is found in Table 10.1. In the table, for students who are bilingual with English as one of their languages, the last three columns would be deleted or left blank. However, for multilingual students these columns may be important. Table 10.1 is different from the flowcharts of language shown in Figures 10.1–10.3. Those flowcharts attempted to show how individual students, following the modelling provided by their teacher, may switch between language forms as they enter into mathematical discussion.

Earlier it was noted that research shows that students who have high competencies in their languages will often academically outperform other students (Clarkson, 2006b; Cummins, 1991, 2000a). Hence, this chart could be used to indicate the languages of the students, but also the level of proficiency they have reached in each language form, within each language. The use of a simple rating system of say 'excellent', 'good' or 'low' in each cell would probably be sufficient. The information will, of course, change over time. Hence updates of the chart should be carried out on a routine basis.

Many have seen the language complexity of classrooms as a difficulty. Trying to map languages is one way of beginning to deal with this complexity and perhaps to see that there can be advantages within the complex nature of this classroom context. At the very least, the teacher will know what they are dealing with and that indeed it is very different from a monolingual context. Moving on from that understanding probably requires that the teacher has some insight into the research briefly sketched in above.

Engaging the student cognitively with their languages

Once a teacher has 'language mapped' their classroom, s/he will have an understanding of what languages are available for use by students and the competencies they have in these languages. As well as beginning to work with the students' communities (the fourth point, below), the teacher could also start working at the individual level in enhancing students' use of their languages when working mathematically. This involves the teacher, and then the teacher and students, using the model in Figure 10.3.

In beginning the use of this strategy, there are at least three issues to keep in mind. Firstly, a student may be using each of their languages intermittently when doing mathematics, for a variety of reasons. Teachers need to be aware of this and also show a positive interest in when and why the student uses their non-English language(s). The second issue deals with the student's proficiencies in their languages. The teacher needs to encourage each student to work with their languages as a useful strategy in solving mathematical problems. This will cue the student to expand their knowledge of that language's mathematical register. Thirdly, the teacher needs to model the way language can be best used in thinking through mathematical ideas and processes, and explicitly encourage students to follow this same process. Beginning with sifting through mathematical ideas in the informal language used by students, using student idioms and their world to situate these ideas, will situate these ideas firmly in the student's space. Since the student's space is in part encoded in their non-English language, it is important for the student to access their space in the language that suits them. But mathematics, even in the lower years, often transcends the informal. For this to occur, language use needs to change. Hence, the teacher will encourage the student to progress through more structured language, to the academic mathematical language appropriate to their year, using their non-English language when they can.

In this process, it is also important for the student to develop their proficiency in English for their mathematical learning. So as well as the teacher encouraging the use of student's first language during mathematics learning, the student also needs to be encouraged at the same time to make the transition in English from the informal through the more mathematical structured to academic mathematical English.

What strategies can teachers use which will encourage both processes? One way to encourage a rich language environment to grow is for teachers to use open-ended questions. These are questions that do not have a specific answer to complete the solution process. They give rise to a multiple number of possible correct solutions (Sullivan & Lilburn, 2002). When learning situations are created around these types of questions, students will need to defend their answers and so to learn how to justify and explain their answers to themselves and others.

In these situations the teacher's role is to engage the students so they are led to a higher level of thinking and discussion. Hence the role is one of a supportive gentle probing inquisitor of the ideas being generated by the students. The teacher gives emotional support and acts to provide a safety net for the students who are engaged together in 'risky thinking', but rarely should the teacher be the provider of answers. This role goes beyond asking students what answers they found, how they obtained them and then giving positive or negative feedback as appropriate. In the type of situation envisaged here, the teacher needs to engage at a deeper level. This is accomplished by careful use of follow-up 'why'-type questions. After actively listening to the students' answers and acknowledging them, the teacher needs to probe further:

- 'Explain that bit to me again';
- 'Does this bit of the answer seem correct to you ... Why?';
- 'Explain that relationship to me again';

or add guidance:

- 'Can you explain that previous example to me please';
- Does that previous problem relate to this one in some way?';
- 'Um could this ... be a legitimate answer for this problem [giving a counter example to that obtain by the student]';

and sometimes reteaching to enable students to progress:

- 'What do you think this [pointing to a term or symbol] means ...?'

and hence eliciting possible meanings of an ambiguous term or proposition. In challenging the students cognitively in this way, undoubtedly the teacher talk, and progressively the student talk, will be using logical connectives (e.g. if, but, however, if ... then, and) as their ideas are built into explanations for their solutions.

The notion of challenging students' thinking with open-ended problems and orchestrating the rich discussion that ensues is no different from the advice that might well be given to teachers in a monolingual teaching context. If the bilingual students are to become competent in English, as well as their first languages when thinking mathematically, then it seems important to keep the same basic structure of this teaching strategy. But it is important to encourage students to make the cross links to their first language at each stage as they move down the flowchart, or indeed back to English if the student has chosen to move to their non-English language. This is probably easily done for the informal language level where students have command of this type of language in both their languages. It may become a little more difficult for some students at the next 'more structured' level, depending on their competence in either or both of their languages. Whether the student has competence in the academic mathematical language in their first language and whether this can be grown is more problematic. However, an exploration of this possibility with the students' communities will give insight into possibilities (see fourth 'engagement strategy' below). Of course for English, this language will need to be taught.

Clearly, the whole process is far more fluid than is depicted above. It is likely that many students use the strategy of switching between their languages when solving mathematical problems in the classroom. However, it seems that not all bilingual students are aware of this behaviour and of how it may be a helpful strategy. As one Year 4 Italian/English bilingual student said to the author during an interview 'I never knew I did that (switched languages). Ye know, no one ever asked me before how I solved maths problems'. The teacher's task in some ways is to slow the thinking processes down a little for students, so students can appreciate how they are processing the solution of the problem by reflecting on the strategies they are using. This would include encouraging bilingual students to reflect on which languages they were using and all students on the type of the language(s) (see Figure 10.3) they were using in that process.

For the teacher, who is probably a monolingual English speaker in the Australian context, encouraging and listening to the students when they use their first languages have to be taken on faith. This may be easier when there is a group of the same non-English speakers in the class. Small group work can from time to time be organised in language groupings, as it will be in friendship and ability groupings at other times. For all bilingual students, but particularly for students who are not part of a wider language group within the class, continuing contact with parents and encouragement for them to support their children's mathematical learning in their first language, should be a teacher priority (see fourth 'engagement strategy' below). Another supporting mechanism can be inviting parents into the mathematics classroom to act as teacher helpers, with the request to use their first language with students frequently. This strategy works even better if the parents are given some preliminary guidelines on how to encourage students to work with open-ended problems.

Engaging the expertise of ESL colleagues

State governments in Australia do provide some assistance to ELL students in developing proficiency in English. Although such help has diminished in the last decade, there is still a network of well-qualified ESL teachers throughout government education sectors and some in the non-government sectors. In large secondary schools that have a significant population of ELL students, ESL teachers may be specific appointments on staff. However, in most Australian primary schools, since they are quite small (200–300 students), the appointment of ESL teachers is much less frequent. But it is with these specifically qualified colleagues that teachers may find help in working creatively with ELL students learning mathematics.

Arkoudis (2003, 2005, 2006), for example, recounts how an ESL teacher worked with a secondary school science teacher who was teaching many ELL students in his science classes in an Australian school. Her account shows clearly the intricacies that are involved and the dedication needed over time as the teacher comes to understand how he needs different emphases within his already effective teaching style. The teacher had to come to grips with more specific attention to his use of language if his students were going to grasp the nuances of science learning he was trying to teach.

If an ESL teaching colleague is not available, then teachers interested in developing their repertoire of teaching strategies could base this on ESL aspects of the curriculum.[3] Although it was noted earlier that there was no mention made of mathematics in the ESL curriculum document, nevertheless, such documents and web sites are very informative as to the teaching of ELL students, and readily accessible.

Engaging the bilingual students' communities

Although this chapter has focused mainly on the cognitive aspects of language use in the classroom, the social and cultural contexts of the students' languages cannot be ignored. In actively supporting the use of a student's non-English language, at a deep level support is given to the student as a social being, something that is crucial to their overall wellbeing, but also for their confidence and willingness to engage in school learning.

In some ways it seems relatively easy to engage students' communities in other areas of schoolwork. If one is to take language studies, then there is obvious overlap. But links to social studies, history, geography and even religion can be made without too much effort. The Multiliteracy Project[4] in Canada has explored the use of immigrant primary students writing dual language books as students seek to articulate and explore their coming to a new country where the teaching language has changed for them. In creating their books students make links to the above areas of the curriculum. But links made to mathematics seems to be more difficult. Perhaps more directed activities that raise the explicit possibilities of mathematical connections are needed; a similar dual language project could be useful for mathematics.

There are some reasonably straightforward issues that could be explored by students. Thomas (1986) compiled a listing of various mathematical symbols and algorithms embedded in different languages that refer to the same mathematical idea or process. Starting with these listings, teachers could ask students to either check them with their parents to see whether the parents had used these when they went to school, or work with their parents to extend the listings to include all the languages represented in a classroom. A comparison of algorithms would lend itself to discussion of efficiency and ease of understanding, both of which are important. More exploration would lead to deeper understandings of historical development of number within other societies, particularly for non-western societies. An exploration of web sites focused on such number work would be fruitful, and mean that for some students, their first language would become the only means by which some sites could be accessed.

Notions of time are also worth exploring. A historical exploration of how people from students' original countries thought of and measured time would be interesting. In western cultures there is a basic notion that time travels forward in a straight line (the arrow of time). In some cultures in Papua New Guinea, notions of long expanses of time are thought of as an arc of a circle, although it is problematic whether there ever is/was a complete circle. Clearly the origins and thinking embodied by the 'Chinese' (and its variants in Southeast Asian countries) calendar would be fruitful to explore with the help of knowledgeable community members.

The use of literature is another possibility. Storybooks have been specifically written for teaching mathematical ideas (e.g. Hutchins, 1970). Folk tales can also be a source of mathematical problems. For example explorations of estimations of times and of distances can be sourced from the tale of Little Red Riding Hood (Clarkson, 2006c). There will be similar tales embedded in the languages of the students that can be thought through from a mathematical viewpoint. Students could set out on such an exploration with the adult members of their communities. It maybe that appropriate tales discovered by the students are only in oral form, and the student(s) will need to translate them into written form. This will mean rich discussion on how to translate mathematical ideas into English.

In each of the above explorations the input from the students' various communities is the key source. Students could create their storybooks, in part at least using digitally recorded interviews with members of their own communities, as explanations and ideas are explored. Again, translation for the whole of their classroom community into English is valued to give access to all. Translation into English does not detract from the power of the first languages being used. Such activities may go some distance in empowering students and their communities.

All parents want to see their children succeed to the best of their abilities. For immigrant parents often the education process is perceived as the avenue that can be used to ensure their children do succeed and mathematics is assumed to be a key component of that success. Hence, it is understandable that these parents will often encourage their children to use English when doing mathematics. Schools need to work hard to convince such parents that the role of their own first language is important for their child's academic progress and they can have an important role in that success. They need to be encouraged to work with their children's teachers who also value the non-English languages brought into their classrooms and actively teach to engage their students mathematically using all languages and modes of language that are available.

Notes

1. The four journals were *Educational Studies in Mathematics, Journal for Research in Mathematics Education, For the Learning of Mathematics* and *Mathematics Education Research Journal.*
2. http://www.qsa.qld.edu.au/qcar/els_files.html#maths.
3. For example http://csf.vcaa.vic.edu.au/es/kses.htm and http://vels.vcaa.vic.edu.au/support/esl/esl.html.
4. See http://www.multiliteracies.ca/index.php.

Chapter 11
Summing Up: Teaching and Learning Mathematics in a Multilingual World

RICHARD BARWELL

Teaching and learning mathematics in multilingual settings is replete with tensions. In Chapter 1, I referred to three tensions apparent in earlier research (notably that of Adler, 2001):

- Tension 1: between mathematics and language;
- Tension 2: between formal and informal language;
- Tension 3: between students' home languages and the official language of schooling.

These three tensions are also apparent in much of the work reported in this book. Tension 1, for example, is apparent in Chval and Khisty's chapter, in the process of drafting and redrafting written mathematical texts. In responding to each new version, the teacher must decide how much to focus on the mathematics and how much to focus on the language the student is using to explain the mathematics. Drafting and redrafting is a fairly long-term process that casts the tension in a new light. Over time, the teacher is able to address both mathematics and language, with a cumulative effect on how the students appear both to understand mathematics and to write about their understanding. The use of writing also adds another dimension. The tension between mathematics and language can be very immediate in the moment-by-moment interaction and discussion orchestrated by the teacher. In working with students through writing mathematics, however, there is more space; through the drafting process, students can return to their writing and think about and work on different aspects at different times. This working across mathematics and language through writing is also apparent in my own chapter, in which the two students, Cynthia and Helena, at different times attend to mathematical and linguistic aspects of their word problem, revising their ideas several times. As I point out, mathematics tasks involving writing seem to promote a productive, reflexive interaction between mathematics and language on the part of the students. Writing tasks, however, may not be the only way to promote this kind of interaction.

Tension 2 is apparent in several chapters, in the guise of a concern with how students come to develop facility with formal mathematical language. Moschkovich, for example, traces in fine detail how two students use a similar phrase 'I went by' to talk about the scales on their graphs. The teacher in Moschkovich's chapter works with the two students to unpack their different interpretations of the scale on the graph and of their informal expression of their ideas. Through this discussion and through the teacher introducing relevant distinctions in order to focus the students' attention, the ideas become more refined and more mathematical. Part of the tension between informal and formal language concerns when and if teachers should intervene to correct students' informal explanations. The teacher in Moschkovich's chapter does not explicitly 'intervene', preferring to work on clarifying different interpretations. Over time, this process is likely to lead to a gradual formalisation of students' mathematical language. Indeed, Clarkson's model proposes that teachers first work with students' 'informal' mathematical talk and, rather like the teacher described by Chval and Khisty, seek to refine this talk over time. Clarkson, however, suggests that all students' languages could be involved in this process.

Tension 3 is particularly evident in some of the settings in which more than one language is used in mathematics classrooms. As Jones, for example, describes, not all students in a Welsh mathematics classroom are necessarily equally proficient in Welsh. How and when the teacher uses Welsh and English is therefore a constant issue. Halai highlights a slightly different aspect of this tension. When the students in her chapter use Urdu to make sense of a problem presented to them in English, they are faced with two possible interpretations for a particular word. These interpretations arise both from drawing on another language (Tension 3) and from the informal nature of that language in the context of a real-world problem (Tension 2). Halai relates one of their interpretations to an incorrect mathematical solution and argues that codeswitching, which has generally been seen as a useful strategy in many situations, needs to be examined more critically.

These three tensions, as originally formulated by Adler (2001) and others (e.g. Khisty, 1995; Moschkovich, 1999a) arise from classroom practice, apparent from careful attention to the development of mathematical meaning and to students' development of suitably formal mathematical language. These tensions are situated within a broader social context, however, that includes local and national language policies, the relative status of different languages and the social status associated with different varieties or styles of language use.

The Social Context of Multilingual Mathematics Classrooms

Language policy, whether at school level or at national level, provides the framework for much classroom practice. Such policies are motivated by a variety of aims, including:

- the maintenance of a particular language, as in Wales;
- the perceived need for students to develop proficiency in a global language like English, as in Malta or in English-medium schools in Pakistan;
- the perceived need for students from linguistic minorities to learn an official or national language, as in Australia, England or the United States.

Where teachers and students share two or more languages with students, as in the first two situations, the ideal is more often than not compromised: on occasions, teachers or students will draw on any of their shared languages to do mathematics. Here are some examples from chapters in this book.

(1) In the classrooms investigated by Farrugia, students used Maltese during group work; more rarely during whole-class talk, of which the following is an example (Farrugia, this volume, transcript modified):

Kirsty: Miss, għaliex għandna tnejn [skali]?
Teacher: English!
Kirsty: Miss, why do we have two, em, x-axis and y-axis (sic)?

(2) In group discussions observed by Halai, students consistently drew on both Urdu and English when working on tasks presented in English (Halai, this volume, transcript modified):

Samina: Iska matlab yeh hai ke Sara 28 years ki hai
Shabnum: Nahi 28 years *ki hogi* because after nine years
Maheen: Because yehan 'will' (hai) 'will' means future
Shabnum: hogi nahi nahi vo hogi nine years ke bad
Maheen: hogi aise batao ke yehan will hai will means future tense

(3) In one example in Jones (this volume, transcript modified), the teacher draws on both Welsh and English in addressing the class. One student then responds in English, another in Welsh:

T: **ond mae ffordd haws ok** there's an easier method of finding twenty five percent
P: find 50% then halve it
T: ok. ok. fifty per cent then halve it
P: **Miss. chwarter e**

In each of these examples, there is a tension (for both students and teachers) between the policy goal and classroom practice. In the first example, by using Maltese, the student appears to have contravened a classroom rule, as evidenced by the teacher's strong response. In the third example, however, the students seem quite free to contribute in either English or Welsh. The second example perhaps falls between the first and the third: the students draw liberally on both Urdu and English, although Halai reports that in whole-class contributions they often sought permission before using Urdu. The tension between language policy and mathematics learning is apparent through comparing these three situations. In the first example, it appears that language policy takes precedence, with the risk that, as Farrugia shows, some students may struggle to fully participate. In the third example, mathematical discussion is facilitated through the use of both Welsh and English, with the risk that at least some students may have less opportunity to develop their proficiency in Welsh. The second example is also interesting, since the use of elements of Urdu and English is described by Halai as a marker of social status: liberal use of English along with Urdu indicates high levels of education and success and may, therefore, implicitly, be a desirable outcome for students.

The three examples discussed above are from multilingual mathematics classrooms in which two or more languages are used by participants. In other settings described in this book, such as Australia, England and the United States, students' home language(s) often have relatively low status and are not heard in mathematics classrooms. In these situations, there is a *de facto* language policy that privileges the societally dominant language. Hence, in England, for example, a Cantonese-speaking immigrant like Cynthia sees little recognition of her language in general or of her own linguistic skills in particular. Cantonese has a relatively low status. The question therefore arises of how, in such settings, students' home language(s) could be harnessed to support their learning of mathematics. This question is not meant to imply that drawing on students' home language(s) will always be advantageous; merely that it may be in some cases, and perhaps more cases than has been realised. Certainly there is evidence that students make more use of their home languages in learning mathematics than their teachers may realise, as is apparent, for example, in the chapters by Monaghan, Chval and Khisty, Staats or Clarkson.

Complicating the Picture

The tensions arising within a multilingual mathematics classroom are, then, situated within a broader social context. This context draws on a set of assumptions or ideologies concerning the nature of language and the nature of mathematics that includes:

- mathematics is language and culture free;
- the language of mathematics is precise and unambiguous;
- there is a standard version of formal mathematical English;
- there are standard versions of natural languages;
- the use of one standard version of a language at a time is the 'normal' and 'correct' way to speak; and consequently
- 'mixing' or 'switching' languages is degenerate.

Of course, these are not the only assumptions about language or mathematics circulating in society. I suggest, however, that the above list form a particular interlocking set of assumptions that informs much discussion and debate about mathematics and language. Combinations of these assumptions inform a number of widespread ideas concerning the role of language in the teaching and learning of mathematics, including:

- the idea that mathematics is the same in all languages and, therefore, that other languages have nothing new or unique to offer mathematics;
- the idea that learning mathematics in an additional language is not a linguistic challenge (because mathematics is language free);
- policies that mandate that only one language be used in mathematics classrooms in multilingual societies;
- the idea that drawing on multiple languages to learn mathematics is problematic or confusing.

The assumptions on which these ideas are based have, however, all been challenged. Recent research in mathematics education has examined, for example, the mutually shaping relationship between language structure and the development of mathematics (e.g. Barton, 2008), the role of ambiguity in mathematical thinking (e.g. Barwell, 2005d) and the varied nature of formal mathematical language in English (e.g. Moschkovich, 2002). Meanwhile, research in applied linguistics has developed perspectives on multilingualism and multilingual societies that recognises a wide range of 'in-between-ness' in language use (see, e.g. Hornberger, 2008; Makoni, 2003). Hence, the boundaries between languages or language varieties are fuzzy and multilingual people use a repertoire of different parts of different languages. Indeed, Makoni argues that, in many multilingual societies, such multiple language repertoires are the 'default' state, although the particular nature of the repertoire will vary from person to person. Such a speaker cannot be reduced to a set of distinct 'one-language-speakers': multiplicity is the norm and the clearly defined, separate use of distinct languages is the degenerate case.

The challenges to these assumptions further complicate the tensions involved in teaching and learning mathematics in multilingual classrooms. Consider the data included in Jones' chapter, for example, in which participants use a mixture of Welsh and English. As Jones points out, this mixture reflects language use in wider Welsh society. While it can be characterised as codeswitching, for the speakers, the 'code' may be the mixture, so that there is less of a sense of 'switching'. It may be, therefore, that the mixture recorded by Jones is an appropriate form of language use for learning mathematics, since it reflects the participants' usual ways of talking. To impede the use of such ways of talking may, in some instances, leave participants with part of their language repertoire unavailable for making mathematical meaning. This is not to say that students do not need to be aware of the nature of mathematical registers in one or more languages (although the interesting question arises of what a mathematical register in a language mixture looks like); clearly, to participate in advanced mathematics, for example, students need to learn how mathematicians communicate. Even mathematicians, however, in the development or communication of their ideas, draw on more informal ways of interacting, including, I conjecture, the use of multiple language repertoires. This observation suggests that Clarkson's model could be further modified to include 'in-between' states.

The chapters by Staats and by Monaghan also helpfully complicate the picture, showing how 'the language of mathematics' is not monolithic. Staats' examination of the metaphors embedded within Somali mathematical language highlights how each language brings the potential for thinking about the supposedly universal ideas of mathematics in a variety of ways. The image of a dependent variable as a passenger on a camel is intriguing. This image also carries an alternative perspective on the concept of dependent variable when compared with English: the camel stands for the function that carries both variables, in contrast to a slightly different relationship implied by the idea of x being a function of y. Thus, Staats' explorations of Somali throw up more than mere linguistic curiosities; there is a potential impact on mathematics. In his chapter, Monaghan looks at the teaching language, rather than the students' language(s). He reveals that the language of mathematics is not the same across the curriculum. A word like 'diagonal' is not used in a standard way; rather, its use varies according to the nature and level of the activity (see also Morgan, 2005). These kinds of findings invite reflection on the idea that, for example, students must learn 'the' formal, standard way of talking about or reading mathematics: what counts as 'standard' or formal is somewhat fuzzy.

A Strategy for Teaching and Learning Mathematics in a Multilingual World

This book illustrates how teaching and learning mathematics in multilingual classrooms situated within increasingly complex multilingual societies is replete with tensions and assumptions. How, then, can we, as teachers, respond? The task is undoubtedly complex and challenging, but students like the ones that appear in the chapters in this book cannot be ignored. It is likely that most mathematics classrooms in most parts of the world are multilingual in the sense used in this book; multilingualism is fast becoming the norm, even in much of the developed world where it has long been marginalised.

I have used the word tension to capture the sense of irresolvable issues. Tensions are not 'either–or' choices; we are not attempting to choose between mathematics and language or between one language and another. Rather, the different parts of these tensions are ever-present, needing to be addressed. The challenge is to find strategies with which to make these tensions productive for learning mathematics and learning language. Of course, as teachers, we seek and develop such strategies through our practice. The chapters in this book have one general strategy in common: the explicit discussion or analysis of language in multilingual mathematics classrooms to inform teaching and learning. The various chapters include, for example,

- analysis of the mathematics register of the classroom language;
- analysis of the nature of the mathematical register in students home language(s);
- attention to and discussion of aspects of the genre of mathematical word problems;
- discussion of students' interpretations of the scale on a graph;
- feedback on students' written mathematics;
- examining language policy documents in relation to mathematics classroom practice;
- collecting and discussing information about students' language backgrounds.

These examples involve an examination of aspects of language by students, teachers or researchers in the context of teaching and learning mathematics. Interestingly, many of these specific examples could be conducted by any of them. For example, Monaghan's analysis of the language of a mathematics teaching programme could be conducted by a group of students as part of a project. Similarly, language policy could be dissected by a group of teachers or even students. In each case, it is the explicit examination of language that is powerful, since it necessarily provokes deeper consideration of mathematical concepts or of mathematics teaching. When Staats interviews Somalis about their mathematical language, both of them learn something about mathematics and something about language. When the students in my chapter argue over the wording of their word problem, they each learn something about mathematics and something about the language used in mathematics classrooms. And when the students in the chapter by Chval and Khisty interact with their teacher through feedback on their mathematical writing, all of them learn something about mathematics and something about language. I do not wish to suggest that a focus on language is likely to be universally useful or applicable. Indeed, one of the three tensions at the start of this chapter concerns the issue of when to shift the focus from mathematics to language (or *vice versa*). It is difficult to make this decision, however, without a sense of what the language issues are. Examining, discussing or analysing language in the context of the mathematics classroom is one useful approach, that, applied in different ways, can inform teachers or students, so that they are able to see all of the tensions in broader perspective. If, therefore, this book has helped you to think more about the nature of multilingualism in your mathematics classroom and its role in the teaching and learning of mathematics, it will have achieved its aim.

References

Abdelnoor, R.E.J. (1979) *A Mathematical Dictionary*. Leeds: Arnold-Wheaton.
Abdi, A. (1998) Education in Somalia: History, destruction, and calls for reconstruction. *Comparative Education* 34 (3), 327–340.
Adetula, L.O. (1989) Solutions of simple word problems by Nigerian children: Language and schooling factors. *Journal for Research in Mathematics Education* 20 (5), 489–497.
Adetula, L.O. (1990) Language factor: Does it affect children's performance on word problems. *Educational Studies in Mathematics* 21 (4), 351–365.
Adler, J. (2001) *Teaching Mathematics in Multilingual Classrooms*. Dordrecht: Kluwer Academic Publishers.
Aitchison, J.W. and Carter, H. (2004) *Spreading the Word*. Talybont, Wales: Y Lolfa.
Alitolppa-Niitamo, A. (2002) The generation in-between: Somali youth and schooling in metropolitan Helsinki. *Intercultural Education* 13 (3), 275–290.
Arkoudis, S. (2003) Teaching English as a second language in science classes: Incommensurate epistemologies? *Language and Education* 17 (3), 161–173.
Arkoudis, S. (2005) Fusing pedagogic horizons: Language and content teaching in the mainstream. *Linguistics and Education* 16 (2), 173–187.
Arkoudis, S. (2006) Negotiating the rough ground between ESL and mainstream teachers. *International Journal of Bilingual Education and Bilingualism* 9 (4), 415–433.
Arthur, J. (2003) "Baro afkaaga hooyo!" A case study of Somali literacy teaching in Liverpool. *International Journal of Bilingual Education and Bilingualism* 6 (3–4), 253–290.
Austin, J. and Howson, A. (1979) Language and mathematical education. *Educational Studies in Mathematics* 10, 161–197.
Baker, C. (1996) *Foundations of Bilingual Education* (2nd edn). Clevedon: Multilingual Matters.
Baker, C. (2001) *Foundations of Bilingual Education and Bilingualism* (3rd edn). Clevedon: Multilingual Matters.
Baker, C. and Jones, S.P. (1998) *Encyclopaedia of Bilingualism and Bilingual Education*. Clevedon: Multilingual Matters.
Bakhtin, M. (1981) *The Dialogic Imagination*. Austin, TX: University of Texas Press.
Ballenger, C. (1997) Social identities, moral narratives, scientific argumentation: Science talk in a bilingual classroom. *Language and Education* 11 (1), 1–14.
Banfi, C. and Day, R. (2004) The evolution of bilingual schools in Argentina. *International Journal of Bilingual Education and Bilingualism* 7 (5), 398–441.
Barton, B. (2008) *The Language of Mathematics: Telling Mathematical Tales*. New York: Springer.

Barton, B., Fairhall, U. and Trinick, T. (1998) Tikanga reo tātai: Issues in the development of a Māori mathematics register. *For the Learning of Mathematics* 18 (1), 3–9.

Bartram, D. (2001) The development of international guidelines on test use: The International Test Commission Project. *International Journal of Testing* 1 (1), 33–54.

Barwell, R. (2002a) Understanding EAL issues in mathematics. In C. Leung (ed.) *Language and Second/Additional Language Issues for School Education: A Reader for Teachers* (pp. 69–80). Watford: NALDIC Publications Group.

Barwell, R. (2002b) Whose words? *Mathematics Teaching* 178, 34–36.

Barwell, R. (2003a) Patterns of attention in the interaction of a primary school mathematics student with English as an additional language. *Educational Studies in Mathematics* 53 (1), 35–59.

Barwell, R. (2003b) Working on word problems. *Mathematics Teaching* 185, 6–8.

Barwell, R. (2004) 'Guessing' in Year 1 mathematics lessons when English is an additional language. *Proceedings of the British Society into Learning Mathematics* 24 (1), 13–18.

Barwell, R. (2005a) Working on arithmetic word problems when English is an additional language. *British Educational Research Journal* 31 (3), 329–348.

Barwell, R. (2005b) Empowerment, EAL and the National Numeracy Strategy. *International Journal of Bilingual Education and Bilingualism* 8 (4), 313–327.

Barwell, R. (2005c) Integrating language and content: Issues from the mathematics classroom. *Linguistics and Education* 16 (2), 205–218.

Barwell, R. (2005d) Ambiguity in the mathematics classroom. *Language and Education* 19 (2), 118–126.

Barwell, R. (2005e) A framework for the comparison of PME research into multilingual mathematics education in different sociolinguistic settings. In H. Chick and J. Vincent (eds) *Proceedings of the 29th Conference of the International Group for the Psychology of Mathematics Education* (Vol. 2, pp. 145–152). Melbourne: International Group for the Psychology of Mathematics Education.

Barwell, R., Barton, B. and Setati, M. (2007) Multilingual issues in mathematics education: Introduction. *Educational Studies in Mathematics* 64 (2), 113–119.

Barwell, R., Leung, C., Morgan, C. and Street, B. (2002) The language dimension of mathematics teaching. *Mathematics Teaching* 180, 12–15.

Bernardo, A.B.I. (1999) Overcoming obstacles to understanding and solving word problems in mathematics. *Educational Psychology* 19 (2), 149–163.

Bialystok, E. (1992) Selective attention in cognitive processing: The bilingual edge. In R.J. Harris (ed.) *Cognitive Processing in Bilinguals* (pp. 501–513). Amsterdam: North Holland.

Bialystok, E. (1994) Analysis and control in the development of second language proficiency. *Studies in Second Language Acquisition* 16, 157–168.

Borasi, R. (1990) The invisible hand in mathematics instruction. In T. Cooney and C. Hirsch (eds) *Teaching and Learning Mathematics in the 1990's* (pp. 174–182). Reston, VA: National Council of Teachers of Mathematics.

Boztepe, E. (2003) Issues in code switching: Competing theories and models. *Working papers in TESOL & Applied Linguistics.* Columbia University Teachers College. On WWW at http://www.tc.columbia.edu/academic//tesol/webjournal. Accessed 20.1.2007.

Brenner, M.E. (1998) Development of mathematical communication in algebra problems solving groups: Focus on language minority students. *Bilingual Research Journal* 22 (3–4), 149–174.

Brincat, J.M. (2006) *Il-Malti: Elf Sena ta' Storja* [*The Maltese Language: A One Thousand Year History*] (2nd edn). Malta: Pubblikazzjonijiet Independenza (PIN).

Brislin, R.W. (1986) The wording and translation of research instruments. In W.J. Lonner and W.B. Berry (eds) *Field Methods in Cross-Cultural Research* (pp. 137–164). London: Sage.

Cameron, L. and Besser, S. (2004) *Writing in English as an Additional Language*. London: DfES.

Camilleri, A. (1995) *Bilingualism in Education: The Maltese Experience*. Heidelberg: Groos.

Camilleri-Grima, A. (2003) "Do as I say, not as I do". Legitimate language in bilingual Malta! In L. Huss, A. Camilleri-Grima and K.A. King (eds) *Transcending Monolingualism: Linguistic Revitalization in Education* (pp. 55–65). Lisse: Swets and Zeitlingee.

Chval, K. (2001) A case study of a teacher who uses calculators to guide her students to successful learning in mathematics. Unpublished doctoral dissertation, University of Illinois at Chicago, IL.

Clarkson, P. (1983) Types of errors made by Papua New Guinean students. *Educational Studies in Mathematics* 14 (4), 353–367.

Clarkson, P. (1991) Language comprehension errors: A further investigation. *Mathematics Education Research Journal* 3 (2), 24–33.

Clarkson, P.C. (1992) Language and mathematics: A comparison of bilingual and monolingual students of mathematics. *Educational Studies in Mathematics* 23 (4), 417–429.

Clarkson, P.C. (1995) Teaching mathematics to NESB students. *Prime Number* 10 (2), 11–12.

Clarkson, P.C. (1996) NESB migrant students studying Mathematics: Vietnamese and Italian students in Melbourne. In L. Puig and A. Gutierrez (eds) *Proceedings of the 20th Conference of the International Group for the Psychology of Mathematics Education* (Vol. 2, pp. 225–232). Valencia, Spain: International Group for the Psychology of Mathematics Education.

Clarkson, P.C. (2004a) Teaching mathematics in multilingual classrooms: The global importance of contexts. In I. Cheong, H. Dhindsa, I. Kyeleve and O. Chukwu (eds) *Globalisation Trends in Science, Mathematics and Technical Education* (pp. 9–23). Brunei Darussalam: Universiti Brunei Darussalam.

Clarkson, P.C. (2004b) Multilingual contexts for teaching mathematics. In M.J. Hoines and A.B. Fuglestad (eds) *Proceedings of the 28th Conference of the International Group for the Psychology of Mathematics Education* (Vol. 1, pp. 236–239). Bergen, Norway: International Group for the Psychology of Mathematics Education.

Clarkson, P.C. (2005) Two perspectives of bilingual students learning mathematics in Australia: A discussion. In D. Hewitt and A. Noyes (eds) *Proceedings of the 6th British Congress of Mathematics Education* (CD format). Warwick: BSRLM.

Clarkson, P.C. (2006a) Multicultural classrooms: Contexts for much mathematics teaching and learning. In F. Favilli (ed.) *Ethnomathematics and Mathematics Education* (pp. 9–16). Pisa, Italy: Tipografia Editrice Pisana.

Clarkson, P.C. (2006b) Australian Vietnamese students learning mathematics: High ability bilinguals and their use of their languages. *Educational Studies in Mathematics* 64, 191–215.

Clarkson, P.C. (2006c) Rhyming and folk tales: Resources for mathematical thinking. *Australian Mathematics Primary Classroom* 11 (4), 18–23.

Clarkson, P.C. (2007) Australian Vietnamese students learning mathematics: High ability bilinguals and their use of their languages. *Educational Studies in Mathematics* 64 (2), 191–215.

Clarkson, P.C. and Dawe, L. (1997) NESB migrant students studying Mathematics: Vietnamese students in Melbourne and Sydney. In E. Pehkonen (ed.) *Proceedings* of the 21st Conference of the International Group for the Psychology of Mathematics Education (Vol. 2, pp. 153–160). Lahte, Finland: International Group for the Psychology of Mathematics Education.

Clarkson, P.C. and Galbraith, P. (1992) Bilingualism and mathematics learning: Another perspective. *Journal for Research in Mathematics Education* 23 (1), 34–44.

Clarkson, P.C. and Indris, N. (2006) Reverting to English to teach mathematics: How are Malaysian teachers and students changing in response to a new language context for learning? *Journal of Science and Mathematics Education in South East Asia* 29 (2), 69–96.

Clements, K. and Del Campo, G. (1987) *Beginning Mathematics*. Melbourne: Catholic Education Office.

Clemson, D. and Clemson, W. (1994) *Mathematics in the Early Years*. London: Routledge.

Cline, T. and Frederickson, N. (eds) (1996) *Curriculum Related Assessment, Cummins and Bilingual Children*. Clevedon: Multilingual Matters.

Cocking, R.R. and Chipman, S. (1988) Conceptual issues related to mathematics achievement of language minority children. In R.R. Cocking and J. Mestre (eds) *Linguistic and Cultural Influences on Learning Mathematics* (pp. 17–46). Hillsdale, NJ: Lawrence Erlbaum Associates.

Cocking, R.R. and Mestre, J. (eds) (1988) *Linguistic and Cultural Influences on Learning Mathematics*. Hillsdale, NJ: Lawrence Erlbaum.

Cooper, B. (1994) Authentic testing in mathematics. *Assessment in Education* 1 (2), 143–166.

Cooper, B. and Dunne, M. (2000) *Assessing Children's Mathematical Knowledge: Social Class, Sex and Problem-solving*. Buckingham: Open University Press.

Cummins, J. (1991) Interdependence of first- and second-language proficiency in bilingual children. In E. Bialystok (ed.) *Language Processing in Bilingual Children* (pp. 70–89). Cambridge: Cambridge University Press.

Cummins, J. (2000a) *Language, Power and Pedagogy: Bilingual Children in the Crossfire*. Clevedon: Multilingual Matters.

Cummins, J. (2000b) "This place nurtures my spirit": Creating contexts of empowerment in linguistically-diverse schools. In R. Phillipson (ed.) *Rights to Language: Equity, Power and Education* (pp. 249–258). Mawah, NJ: Lawrence Erlbaum Associates.

Cummins, J. (2001) *An Introductory Reader to the Writings of Jim Cummins* (eds C. Baker and N.H. Hornberger). Clevedon: Multilingual Matters.

Dawe, L. (1983) Bilingualism and mathematical reasoning in English as a second language. *Educational Studies in Mathematics* 14 (4), 325–353.

Department for Children, Schools and Families (DCSF) (2007) Schools and pupils in England, January 2007 (Final). On WWW at http://www.dfes.gov.uk/rsgateway/DB/SFR/s000744/index.shtml. Accessed 22.3.2008.

Department for Education and Employment (DfEE) (1999) *The National Numeracy Strategy: Framework for Teaching Mathematics from Reception to Year 6*. Sudbury: DfEE Publications.

Department of Education and Science (DES) (1982) *Mathematics Counts* (The Cockcroft Report). London: HMSO.

Duran, R. (1987) Factors affecting development of second language literacy. In S. Goldman and H. Trueba (eds) *Becoming Literate in English as a Second Language* (pp. 33–55). Norwood, NJ: Ablex.

Edwards, D. (1997) *Discourse and Cognition*. London: Sage.

Edwards, D. and Potter, J. (1992) *Discursive Psychology*. London: Sage.

Ellerton, N. and Clarkson, P.C. (1996) Language factors in mathematics teaching and learning. In A. Bishop, K. Clements, C. Keitel, J. Kilpatrick and C. Laborde (eds) *International Handbook of Mathematics Education* (pp. 991–1038). Dordrecht: Kluwer Academic Publishers.

Ellerton, N., Clements, M.A. and Clarkson, P.C. (2000) Language factors in mathematics education. In K. Owens and J. Mousley (eds) *Research in Mathematics Education in Australasia 1996–1999* (pp. 29–96). Sydney: Mathematics Education Research Groups of Australasia.

Emblen, V. (1988) Asian children in schools. In D. Pimm (ed.) *Mathematics, Teachers and Children* (pp. 82–94). Sevenoaks: Hodder and Stoughton.

Evans, S. (2007) Differential performance of items in mathematics assessment materials for 7-year-old pupils in English-medium and Welsh-medium versions. *Educational Studies in Mathematics* 64 (2), 145–168.

Evans, Z. and Hughes, T. (2003) Yr athro ail iaith yn yr ysgol gynradd. In G. Roberts and C. Williams (eds) *Addysg Gymraeg addysg Gymreig* (pp. 312–336). Bangor: University of Wales Bangor.

Faltis, C. (1996) Learning to teach content bilingually in a middle school bilingual classroom. *The Bilingual Research Journal* 20 (1), 29–44.

Farrugia, M.T. (2007) Medium and message: The use and development of an English Mathematics register in two Maltese primary classrooms. Unpublished doctoral dissertation, University of Birmingham.

Favilli, F. and Jama Musse, J. (1996) Creating a mathematical terminology: The Somalia case. Paper presented at the *8th International Congress on Mathematical Education*, Sevilla, Spain, July.

Franceschini, R. (1998) The notion of code in linguistics. In P. Auer (ed.) *Code-switching in Conversation: Language, Interaction and Identity* (pp. 51–72). London: Routledge.

Garcia, E. (2001) *The Education of Hispanics in the United States: Raíces y Alas*. Lanham, MD: Rowman & Littlefield Publishers, Inc.

Gee, J. (1999) *An Introduction to Discourse Analysis: Theory and Method*. New York: Routledge.

Gerofsky, S. (1996) A linguistic and narrative view of word problems in mathematics education. *For the Learning of Mathematics* 16 (2), 36–45.

Glaser, B. and Strauss, A. (1967) *The Discovery of Grounded Theory: Strategies for Qualitative Research*. Hawthorne, NY: Aldine.

Gorgorió, N. and Planas, N. (2005) Reconstructing norms. In H.L. Chick and J.L. Vincent (eds) *Proceedings of the 29th Meeting of the International Group for the Psychology of Mathematics Education* (Vol. 3, pp. 65–72). Melbourne: International Group for the Psychology of Mathematics Education.

Griffiths, R. and Clyne, M. (1994) *Language in the Mathematics Classroom: Talking, Representing, Recording*. Melbourne: Eleanor Curtain Publishing.

Grosjean, F. (1999) Individual bilingualism. In B. Spolsky (ed.) *Concise Encyclopedia of Educational Linguistics* (pp. 284–290). London: Elsevier.

Gumperz, J.J. (1982) *Discourse Strategies*. Cambridge: Cambridge University Press.

Gutstein, E. (2007) Multiple language use and mathematics: Politicizing the discussion. *Educational Studies in Mathematics* 64 (2), 243–246.

Haan Associates (1992) *Study Companion Word List: Mathematics/Eraybixinta: Xisaab*.

Halai, A. (2001) Role of social interactions in students' learning of mathematics (in classrooms in Pakistan). Unpublished doctoral dissertation: University of Oxford.

Halai, A. (2006) Mentoring in-service teachers: Issues of role diversity. *Teaching and Teacher Education* 22 (6), 700–710.

Halai, A. (2007) Learning mathematics in English medium classrooms in Pakistan: Implications for policy and practice. *Bulletin of Educational Research* 29 (1), 1–15.

Halcón, L., Robertson, C., Savik, K., Johnson, D., Spring, M., Butcher, J., Westermeyer, J. and Jaranson, J. (2004) Trauma and coping in Somali and Oromo youth. *Journal of Adolescent Health* 35, 17–24.

Halliday, M.A.K. (1978) *Language as Social Semiotic: The Social Interpretation of Language and Meaning*. London: Edward Arnold.

Halliday, M.A.K. and Martin, J.R. (1993) *Writing Science: Literacy and Discursive Power*. London: The Falmer Press.

Hambleton, R.K. (1993) Translating achievement tests for use in cross-national studies. *European Journal of Psychological Assessment* 9 (1) 57–68.

Haque, A.R. (1993) The position and status of English in Pakistan. In R. Baumgardner (ed.) *The English Language in Pakistan* (pp. 13–18). Karachi: Oxford University Press.

Hargreaves, E. (1997) Mathematics assessment for children with English as an additional language. *Assessment in Education* 4 (3), 401–411.

Hart, K.M. (1981) *Children's Understanding of Mathematics: 11–16*. London: John Murray.

Harvey, R. (1982) 'I can keep going up if I want to': One way of looking at learning in mathematics. In R. Harvey, D. Kerslake, H. Shuard and M. Torbe (eds) *Mathematics* (pp. 22–40). London: Ward Lock Educational.

Hasan Aly, J. (2006) *Education in Pakistan: A White Paper*. Prepared with the National Educational Policy Review Team. Islamabad, Pakistan: Government of Pakistan Ministry of Education.

Heath, S.B. (1983) *Ways with Words: Language, Life, and Work in Communities and Classrooms*. Cambridge: Cambridge University Press.

Heller, M. (1999) *Linguistic Minorities and Modernity*. London: Longman.

Heller, M. and Martin-Jones, M. (eds) (2001) *Voices of Authority: Education and Linguistic Difference*. Westport, CT: Ablex Publishing.

Hornberger, N. (2008) Continua of biliteracy. In A. Creese, P. Martin and N.H. Hornberger (eds) *Encyclopedia of Language and Education* (2nd edn), Vol. 9: Ecology of Language (pp. 275–290). New York: Springer.

Howie, S.J. (2002) English language proficiency and contextual factors influencing mathematics achievement of secondary school pupils in South Africa. Doctoral dissertation, University of Twente.

Howie, S.J. (2003) Language and other background factors affecting secondary pupils' performance in mathematics in South Africa. *African Journal of Research in Mathematics, Science and Technology Education* 7, 1–20.

Hunston, S. (2002) *Corpora in Applied Linguistics*. Cambridge: Cambridge University Press.

Hutchins, P. (1970) *Clocks and More Clocks*. New York: Aladdin.

International Test Commission (2000) *International Guidelines for Test Use*. Louvain-la-Neuve, Belgium: ITC.

Jacobson, R. (1990) Allocating two languages as a key feature of bilingual methodology. In R. Jacobson and C. Faltis (eds) *Language Distribution Issues in Bilingual Schooling* (pp. 3–17). Clevedon: Multilingual Matters.

Jama Musse, J. (1998) The role of ethnomathematics in mathematics education: Cases from the Horn of Africa. Paper presented at the First International Conference on Ethnomathematics, Granada, Spain, September. On WWW at http://www.fiz-karlsruhe.de/fiz/publications/zdm/zdm993a2.pdf. Accessed 9.1.2007.

Johnson, J. (2006) Orality, literacy, and Somali oral poetry. *Journal of African Cultural Studies* 18 (1), 119–136.
Jones, D.V. (1993) Words with a similar meaning. *Mathematics Teaching* 145, 14–15.
Jones, D.V. (1998) National curriculum tests for mathematics in English and Welsh: Creating matched assessments. *Assessment in Education* 5 (2), 193–211.
Jones, D.V. (2000) Talk and texts in bilingual mathematics lessons in Wales. *The Welsh Journal of Education* 9 (2) 102–119.
Jones, D.V. and Martin-Jones, M. (2004) Bilingual education and language revitalization in Wales: Past achievements and current issues. In J. Tollefson and A. Tsui (eds) *Medium of Instruction Policies: Which Agenda? Whose Agenda?* (pp. 43–70). New Jersey: Lawrence Erlbaum Associates.
Karplus, R., Karplus, E., Formisano, M. and Paulsen, A.C. (1979) Proportional reasoning and control of variables in seven countries. In J. Lochheed and J. Clement (eds) *Cognitive Process Instruction*. Philadelphia, PA: The Franklin Institute Press.
Keenan, P.J. (1879) *An Inquiry Into the System of Education in Malta*. London: Charles Thom.
Khan, A.R. (2002) Language and content: A case study of how science is taught in a community run private English medium school in Karachi Pakistan. Unpublished masters thesis, Ontario Institute for Studies in Education, University of Toronto.
Khan, M.A. and Khan, M.A. (2002) Writing Urdu in Roman. *The Dawn Daily*, 17th November.
Khisty, L.L. (1995) Making inequality: Issues of language and meaning in mathematics teaching with Hispanic students. In W. Secada, E. Fennema and L.B. Adajian (eds) *New Directions for Equity in Mathematics Education* (pp. 279–297). Cambridge, UK: Cambridge University Press.
Khisty, L.L. (1996) Children talking mathematically in multilingual classrooms: Issues in the role of language. In H. Mansfield, N. Pateman and N. Bednarz (eds) *Mathematics for Tomorrow's Young Children: International Perspectives on Curriculum* (pp. 240–247). Boston, MA: Kluwer Academic Publishers.
Khisty, L.L. (2001) Effective teachers of second language learners in mathematics. In M. van den Heuvel-Panhuizen (ed.) *Proceedings of the 25th Conference of the International Group for the Psychology of Mathematics Education* (Vol. 3, pp. 225–232). Utrecht: Utrecht University.
Khisty, L.L. and Chval, K. (2002) Pedagogic discourse and equity in mathematics: When teachers' talk matters. *Mathematics Education Research Journal* 14, 154–168.
Kress, G. and van Leeuwen, T. (1994) *Reading Images: The Design of Visual Communication*. London: Routledge.
Lakoff, G. and Núñez, R. (2000) *Where Mathematics Comes From: How the Embodied Mind Brings Mathematics into Being*. New York: Basic Books.
Lamon, S. (1994) Ratio and proportion: Cognitive foundations in unitizing and norming. In G. Harel and J. Confrey (eds) *The Development of Multiplicative Reasoning in the Learning of Mathematics* (pp. 89–120). New York: Suny Press.
Lamon, S. (1996) The development of unitizing: Its role in children's partitioning strategies. *Journal for Research in Mathematics Education* 27 (2), 170–193.
Laporte, R. (1998) Pakistan. *World Book Multimedia Encyclopedia* (CD Rom). Chicago, IL: World Book Inc.
Lappan, G., Fey, J.T., Fitzgerald, W.M., Freil, S.N. and Phillips, E.D. (1998) *Connected Mathematics*. White Plains, NY: Dale Seymour Publications.
Lave, J. (1992) Word problems: A microcosm of theories of learning. In P. Light and G. Butterworth (eds) *Context and Cognition: Ways of Learning and Knowing* (pp. 74–92). Hemel Hempstead: Harvester Wheatsheaf.

Lee, C. (2006) *Language for Learning Mathematics*. Maidenhead: Open University Press.

Leung, C. (2005) Mathematical vocabulary: Fixers of knowledge or points of exploration? *Language and Education* 19 (2), 127–135.

Lindholm-Leary, K.J. (2001) *Dual Language Education*. Clevedon: Multilingual Matters.

Lodholz, R. (1990) The transition from arithmetic to algebra. In E.L. Edwards (ed.) *Algebra for Everyone* (pp. 24–33). Reston, VA: National Council of Teachers of Mathematics.

Makoni, S. (2003) From misinvention to disinvention of language: Multilingualism and the South African constitution. In S. Makoni, G. Smitherman, A.F. Ball and A.K. Spears (eds) *Black Linguistics: Language, Society, and Politics in Africa and the Americas*. London: Routledge.

Martin-Jones, M. (2000) Bilingual classroom interaction: A review of recent research. *Language Teaching* 33, 1–9.

Mehan, H. (1979) *Learning Lessons: Social Organization in the Classroom*. Cambridge, MA: Harvard University Press.

Mejia, A. (2004) Bilingual education in Columbia: Towards an integrated perspective. *International Journal of Bilingual Education and Bilingualism* 7 (5), 381–397.

Mendes, J.R. (2007) Numeracy and literacy in a bilingual context: Indigenous teachers education in Brazil. *Educational Studies in Mathematics* 64 (2), 217–230.

Mercer, N. and Sams, C. (2006) Teaching children how to use language to solve maths problems. *Language and Education* 20 (6), 507–528.

Mestre, J. (1986) Teaching problem-solving strategies to bilingual students: What do research results tell us? *International Journal of Mathematical Education in Science and Technology* 17 (4), 393–401.

Mestre, J. (1988) The role of language comprehension in mathematics and problem solving. In R.R. Cocking and J. Mestre (eds) *Linguistic and Cultural Influences on Learning Mathematics* (pp. 201–220). Hillsdale, NJ: Lawrence Erlbaum.

Miller, L.D. (1993) Making the connection with language. *The Arithmetic Teacher* 41 (2), 311–316.

Ministry of Education (1999) *Creating the Future Together: National Minimum Curriculum*. Malta: Ministry of Education.

Ministry of Education (2006) *National Curriculum for Mathematics Grades I–XII, 2006*. Islamabad, Pakistan: Government of Pakistan Ministry of Education.

Ministry of Education and National Culture (1998) *Kurrikulu Nazzjonali Ġdid għall-Edukazzjoni f'Malta bejn l-eta ta' 3 u 16-il Sena [A New National Curriculum for Education in Malta 3–16]*. Malta: Ministry of Education and National Culture.

Ministry of Education, Youth and Employment (in collaboration with the University of Malta) (2005) *National Conference 2004: The Teaching and Learning of Mathematics in Malta*. Malta: Department for Curriculum Management, Ministry of Education, Youth and Employment.

Mohan, B. (1990) LEP students and the integration of language and content: Knowledge structures and tasks. In J. Gomez (ed.) *Proceedings of the First Research Symposium on Limited English Proficient Student Issues* (pp. 113–160). Washington, DC: Office for Bilingual Education and Minority Language Affairs.

Mohan, B. and Slater, T. (2005) A functional perspective on the critical 'theory/practice' relation in teaching language and science. *Linguistics and Education* 16, 151–172.

Moll, L. (ed.) (1990) *Vygotsky and Education: Instructional Implications and Applications of Sociohistorical Psychology*. New York: Cambridge University Press.

Monaghan, F. (1997) Language and mathematics: The analysis of written text in the secondary classroom. Unpublished doctoral dissertation, University of Reading.

Monaghan, F. (1999) Judging a word by the company it keeps: The use of concordancing software to explore aspects of the mathematics register. *Language and Education* 13 (1), 59–70.
Monaghan, F. (2000) What difference does it make? Children's views of the differences between some quadrilaterals. *Educational Studies in Mathematics* 42 (2), 179–196.
Monaghan, F. (2005) Don't think in your head, think aloud—using ICT to promote collaborative thinking in mathematics. *Research in Mathematics Education* 7, 83–100.
Morgan, C. (1998) *Writing Mathematically: The Discourse of Investigation*. London: Falmer Press.
Morgan, C. (2005) Words, definitions and concepts in discourses of mathematics, teaching and learning. *Language and Education* 19 (2), 103–117.
Morgan, C. (2007) Who is not multilingual now? *Educational Studies in Mathematics* 64 (2), 239–242.
Moschkovich, J. (1999a) Supporting the participation of English language learners in mathematical discussions. *For the Learning of Mathematics* 19 (1), 11–19.
Moschkovich, J. (1999b) Understanding the needs of Latino students in reform-oriented mathematics classrooms. In L. Ortiz-Franco, N. Hernandez and Y. De La Cruz (eds) *Changing the Faces of Mathematics: Perspectives on Latinos and Latinas* (pp. 5–12). Reston, VA: National Council of Teachers of Mathematics.
Moschkovich, J. (2002) A situated and sociocultural perspective on bilingual mathematics learners. *Mathematical Thinking and Learning* 4 (2–3), 189–212.
Moschkovich, J. (2007) Using two languages when learning mathematics. *Educational Studies in Mathematics* 64 (2), 121–144.
Moschkovich, J. (2008) "I went by twos, he went by one:" Multiple interpretations of inscriptions as resources for mathematical discussions. *The Journal of the Learning Sciences* 17 (4), 551–587.
Mousley, J. and Marks, G. (1991) *Discourses in Mathematics*. Geelong: Deakin University.
National Center for Educational Statistics (NCES) (2002) *The Conditions of Education 2002* (2002-025). Washington, DC: U.S. Department of Education.
National Council of Teachers of Mathematics (NCTM) (1989) *Curriculum and Evaluation Standards*. Reston, VA: NCTM.
National Council of Teachers of Mathematics (NCTM) (2000) *Principles and Standards for School Mathematics*. Reston, VA: NCTM.
Noelting, G. (1980) The development of proportional reasoning and the ratio concept. Part I—Differentiation of stages. *Educational Studies in Mathematics* 11, 217–253.
Núñez, R. (2006) Do real numbers really move? In R. Hersh (ed.) *18 Unconventional Essays on the Nature of Mathematics* (pp. 160–181). New York: Springer.
Oakland, T. (2004) Use of educational and psychological tests internationally. *Applied Psychology: An International Review* 53 (2), 157–172.
O'Halloran, K.L. (1998) Classroom discourse in mathematics: A multisemiotic analysis. *Linguistics and Education* 10 (3), 359–388.
Parvanehnezhad, Z. and Clarkson, P.C. (2008) Iranian bilingual students' reported use of language switching when doing mathematics. *Mathematics Education Research Journal* 20 (1), 52–81.
Peregoy, S. and Boyle, O. (1993) *Reading, Writing, and Learning in ESL*. White Plains, NY: Longman.
Phillips, C.J. and Birrell, H.V. (1994) Number learning of Asian pupils in English primary schools. *Educational Research* 36 (1), 51–62.
Pike, N.V. (2004) Co-constructing mathematical knowledge through talk in three Welsh language primary classrooms in South Wales. Unpublished doctoral dissertation, University of Wales.

Pimm, D. (1987) *Speaking Mathematically: Communication in Mathematics Classrooms*. London: Routledge and Kegan Paul.
Pimm, D. (1995) *Symbols and Meanings in School Mathematics*. London: Routledge.
Pujolar, J. (2000) *Gender, Heteroglossia and Power*. Berlin: Mouton de Gruyter.
QCA (UK Qualifications and Curriculum Authority) (1998) Year 5 mathematics test: Mathematics booklet. Sudbury: QCA Publications.
Queensland Studies Authority (2004) *Mathematics: Years 1 to 10 Syllabus*. Brisbane: Queensland Studies Authority.
Quinn, R. and Wilson, M. (1997) Writing in the mathematics classroom: Teacher beliefs and practices. *The Clearinghouse* 71 (1), 14–20.
Rahman, T. (2002) *Language Ideology and Power, Language Learning Among the Muslims of Pakistan and North India*. Karachi: Oxford University Press.
Roberts, G. (2000) Bilingualism and number in Wales. *International Journal of Bilingual Education and Bilingualism* 3 (1), 44–56.
Roberts, T. (1998) Mathematical registers in Aboriginal languages. *For the Learning of Mathematics* 18 (1), 10–16.
Robillos, M. (2001) Somali needs assessment project: A report prepared for the Somali Resource Center. Minneapolis, Minnesota: Center for Urban and Regional Affairs, University of Minnesota.
Rothman, R.W. and Cohen, J. (1989) The language of math needs to be taught. *Academic Therapy* 25 (2), 133–142.
Schwartzman, S. (1994) *The Words of Mathematics: An Etymological Dictionary of Mathematical Terms used in English*. Washington, DC: Mathematical Association of America.
Secada, W.G. (1991) Degree of bilingualism and arithmetic problem solving in Hispanic first graders. *The Elementary School Journal* 92 (2), 213–231.
Secada, W.G. (1992) Race, ethnicity, social class, language and achievement in mathematics. In D.A. Grouws (ed.) *Handbook of Research on Mathematics Teaching and Learning* (pp. 623–660). New York: MacMillan.
Setati, M. (2002) Researching mathematics education and language in multilingual South Africa. *The Mathematics Educator* 12 (2), 6–20.
Setati, M. (2005a) Teaching mathematics in a primary multilingual classroom. *Journal for Research in Mathematics Education* 36 (5), 447–466.
Setati, M. (2005b) Power and access in multilingual mathematics classrooms. In M. Goos, C. Kanes and R. Brown (eds) *Proceedings of the 4th International Mathematics Education and Society Conference* (pp. 7–18). Gold Coast: Centre for Learning Research, Griffith University.
Setati, M. and Adler, J. (2000) Between languages and discourses: Language practices in primary multilingual mathematics classrooms in South Africa. *Educational Studies in Mathematics* 43 (3), 243–269.
Sfard, A. (2000) Symbolizing mathematical reality into being—Or how mathematical discourse and mathematical objects create each other. In P. Cobb, E. Yackel and K. McClain (eds) *Symbolizing and Communicating in Mathematics Classrooms: Perspectives on Discourse, Tools, and Instructional Design* (pp. 37–98). Mahwah, NJ: Lawrence Erlbaum.
Sfard, A., Nesher, P., Streefland, F., Cobb, P. and Mason, J. (1998) Learning mathematics through conversation: Is it as good as they say? *For the Learning of Mathematics* 18 (1), 41–51.
Sierpinska, A. (2002) Book review: Teaching mathematics in multilingual classrooms. *Zentralblatt für Didaktik der Mathematik* 34 (3), 98–103.
Silver, E.A., Kilpatrick, J. and Schlesinger, B. (1990) *Thinking Through Mathematics: Fostering Inquiry and Communication in Mathematics Classrooms*. New York: College Board Publications.

Sinclair, J. and Coulthard, M. (1975) *Towards an Analysis of Discourse: The English Used by Teachers and Pupils*. Oxford: Oxford University Press.

Sinclair, J.M. and Renouf, A. (1988) A lexical syllabus for language learning. In R. Carter and M. McCarthy (eds) *Vocabulary and Language Teaching* (pp. 140–160). London: Longman.

Social Policy Development Centre (SPDC) (2003) *Social Development in Pakistan: The State of Education. Annual Review 2002–03*. Karachi, Pakistan: SPDC.

Sowder, J., Armstrong, B., Lamon, S., Simon, M., Sowder, L. and Thompson, A. (1998) Educating teachers to teach multiplicative structures in the middle grades. *Journal of Mathematics Teacher Education* 1, 127–155.

Stodolsky, S. (1988) *The Subject Matters: Classroom Activity in Math and Social Studies*. Chicago: University of Chicago Press.

Street, B. (2006) So what about multimodal numeracies? In K. Pahl and J. Rowsell (eds) *Travel Notes from the New Literacy Studies: Instances of Practice* (pp. 219–233). Clevedon: Multilingual Matters.

Sullivan, P. and Lilburn, P. (2002) *Good Questions for Math Teaching: Why Ask Them and What to Ask, K–6*. Sausalito, CA: Math Solution Publications.

Swain, M. and Lapkin, S. (1995) Problems in output and the cognitive processes they generate: A step towards second language learning. *Applied Linguistics* 16 (3), 371–391.

Thomas, J. (1986) *Number ≠ Maths*. Melbourne: Child Migrant Education Services.

Thomas, W.P. and Collier, V. (1997) *School Effectiveness for Language Minority Students*. Washington, DC: National Clearinghouse for Bilingual Education.

Thompson, A., Philipp, R., Thompson, P. and Boyd, B. (1994) Calculational and conceptual orientations in teaching mathematics. In D. Aichele and A. Coxford (eds) *Professional Development for Teachers of Mathematics* (pp. 79–92). Reston, VA: National Council of Teachers of Mathematics.

Tollefson, J.W. and Tsui, A.B.M. (eds) (2004) *Medium of Instruction Policies: Which Agenda? Whose Agenda?* Mahwah, NJ: Lawrence Erlbaum Associates.

UNESCO (1975) *Interactions Between Language and Mathematical Education: Final Report of the Symposium* Sponsored by UNESCO, CEDO and ICMI, Nairobi, Kenya, September 1–11, *1974* UNESCO Report No. ED-74/CONF. 808, Paris: UNESCO.

Verschaffel, L., Greer, B. and de Corte, E. (2000) *Making Sense of Word Problems*. Lisse: Swets and Zeitlinger.

Victorian Curriculum and Assessment Authority (2002) *Mathematics Curriculum and Standards Framework 2*. Melbourne: Victorian Curriculum and Assessment Authority.

Vygotsky, L. (1986) *Thought and Language*. Cambridge, MA: The MIT Press.

Warren, B. and Rosebery, A. (1996) This question is just too, too easy. In L. Schauble and R. Glazer (eds) *Innovations in Learning: New Environments for Education* (pp. 97–125). Hillsdale, NJ: Lawrence Erlbaum Associates.

Warwick, D.P. and Reimers, F. (1995) *Hope or Despair? Learning in Pakistan's Primary Schools*. London: Praeger Publishers.

Waters, T. and LeBlanc, K. (2005) Refugees and education: Mass public schooling without a nation-state. *Comparative Education* 49 (2), 129–147.

Welsh Assembly Government (2007a) *Schools in Wales: General Statistics 2006*. Cardiff: Welsh Assembly Government.

Welsh Assembly Government (2007b) *Welsh in Schools 2007*. Cardiff: Welsh Assembly Government.

Welsh Language Board (2003) *Census 2001: Main Statistics about Welsh*. Cardiff: Bwrdd yr Iaith.

Williams, C. (1995) *Dysgu yn y sefyllfa Ddwyieithog*. Cyfres Datblygu Dulliau Addysgu. Bangor: Y Ganolfan Adnoddau.

Williams, C. (1997) *Bilingual Teaching in Further Education: Taking Stock*. Bangor, Wales: Canolfan Bedwyr.

Williams, C., Lewis, G. and Baker, C. (1996) *The Language Policy: Taking Stock*. Caernarfon, Wales: Gwynedd County Council.

Willis, D. (1990) *The Lexical Syllabus: A New Approach to Language Teaching*. London: Harper Collins.

Young, B. (2002) *Characteristics of the 100 Largest Public Elementary and Secondary School Districts in the United States: 2000–01*, NCES 2002–351. Washington, DC: U.S. Department of Education, NCES.

Zaskis, R. (2000) Using code-switching as a tool for learning mathematical language. *For the Learning of Mathematics* 20 (3), 38–43.

Index

Abdelnoor, R. 23
Abdi, A. 34, 36
about 26, 27
academic language 134, 143, 147, 148, 149, 155
Adetula, L. 64, 65
Adler, J. 44, 49, 59, 102, 104, 105, 115, 120, 126, 151, 161, 162
Aitchison, J. 113
algebra 18, 19, 20, 32, 33, 39, 44, 65
algebraic notation 10, 44, 119
Alitolppa-Niitamo, A. 37, 38
ambiguity 7, 19, 41, 92, 93, 156, 165
angles 19, 22, 35, 42-43, 101, 109, 125
anthropological methods 10, 33
applied linguistics 10, 165
appropriation 133, 139
Arabic 16, 34
area (mathematical) 22, 108, 129, 136-141
arithmetic 4, 8, 35, 44, 67, 76
arithmetic word problems (*see* mathematical word problems)
Arkoudis, S. 158
Arthur, J. 38
astronomy 40, 41, 42
Austin, J. 146
Australia 5, 12, 111, 145-160, 163, 164
axes 11, 25, 78, 83-87, 90, 94, 95

Baker, C. 100, 101, 102, 114, 118, 120
Bakhtin, M. 133, 134
Ballenger, C. 79
Banfi, C. 120
barabar 53, 54, 56
Barton, B. 34, 111, 165
Barwell, R. 15, 44, 50, 64, 67, 69, 71, 74, 75, 77, 120, 121, 126, 151, 165
Bengali 28
Bernardo, A. 64
Besser, S. 27
Bialystok, E. 6
bil 40-42
bilingual education 3, 50, 113, 114, 126
bilingual learners 20, 25, 28, 30, 63-77, 78-79, 96, 118, 123, 124, 125, 127, 128, 146, 156-158
bilingual mathematics classroom(s)
– Maltese-English 98-112
– Spanish-English 7, 9, 79-96, 115, 118, 128-144

– Welsh-English 114-127
bilingual pupils (*see* bilingual learners)
bilingual schools 114-126
bilingual students (*see* bilingual learners)
biology 100, 101
Birrell, H. 3
Borasi, R. 79
Boyle, O. 133, 134
Boztepe, E. 49
Brenner, M. 133
Brincat, J. 99
Brislin, R. 124
British National Corpus 20
but 5

Cameron, L. 27
Camilleri, A. 99, 100, 102
Camilleri-Grima, A. 97, 99, 100
Cantonese 63, 164
Carter, H. 113
Chipman, S. 3
Chval, K. 12, 129, 133, 144, 161, 162, 164, 167
Clarkson, P. 5, 6, 12, 44, 49, 50, 59, 64, 65, 110, 126, 145, 146, 150, 151, 152, 154, 159, 162, 164, 166
Clements, K. 147
Clemson, D. 108
Clemson, W. 108
Cline, T. 123
Clyne, M. 107
Cobuild Corpus 9, 21, 23
Cockcroft Report, The 107
Cocking, R. 3, 146
code-mixing 100
codeswitching 48-49, 58-62, 100-111, 120-121
coefficient 44
coercive power structures 50, 61
cognition 6, 50, 62
cognitive advantages 5, 6, 133
cognitive disadvantages 4
Cohen, J. 107
Collier, V. 64
communicative language 145, 146
communicative resources 102, 121, 125
community-based language research 33, 35, 38, 45, 46
comprehensible input 133
conceptual discourse 8
concordancer 10, 15, 20-23, 30

concurrent language approach 118, 119, 120
Confederation of Somali Communities (Minnesota) 38
congruence 129-141
contextualisation cues 121
Cooper, B. 11, 36, 43, 65, 103
cooperative learning 11, 104, 111
corpus analysis 10, 15, 31
corpus linguistics 10
Coulthard, M. 103
crescent moon 10, 34, 41-42
cultural knowledge 38, 42, 43
cultural tool 134
culture 3, 4, 10, 30, 31, 33, 34, 36, 38-46, 50, 60, 64, 100, 113, 124, 126, 127, 135, 141, 144, 145, 159, 165
Cummins, J. 4, 5, 50, 60, 146, 154
curriculum design 15, 32
curriculum-related assessment 123
curves 84, 85, 93, 94
cyflun 125

data representation 39
Dawe, L. 5, 30, 146
Day, R. 120
Del Campo, G. 147
Department for Children, Schools and Families (DCSF) (UK) 15
Department for Education and Employment (DfEE) (UK) 67
Department of Education and Science (DES) (UK) 107
derivative 35
deunaw 124
dheeli 33, 40, 43
diagonal 21-24, 27, 166
diagrams 26, 30, 109, 119
dialogic teaching 16, 135, 144
dictionaries 18-19, 23, 39, 110, 133, 142
discourse(s) 15, 41, 58, 60, 120, 138, 140, 150 (*see also* mathematical discourse)
discourse analysis 120
discursive patterns 120
discursive psychology 68
disempowerment 8
diversity 1, 2, 16, 17, 31, 46, 47, 50, 76, 113, 118, 123
drafting 12, 124, 129-142, 161
Dunne, M. 65
Duran, R. 133

EAL (*see* English as an additional language)
EAL/ESL learners 3, 4, 7, 16, 19, 67, 143
EAL/ESL teachers 19, 20, 153, 157, 158
economics (school subject) 100, 101
Edwards, D. 68, 70
eighteen 124
either...or 15
ELL (*see* English language learners)

Ellerton, N. 146
Emblen, V. 8
empowerment 8, 17, 50, 159
England 11, 14-31, 63-77, 115, 123, 163, 164
English 1, 2, 3, 4, 5, 8, 9, 10, 11, 12, 14, 15, 16, 17, 18, 19, 20, 25, 28, 34, 35, 39, 40, 42, 45, 46, 47, 48, 50, 51, 52, 53, 58, 59, 60, 61, 62, 63, 64, 65, 67, 73, 74, 77, 79, 80, 96, 97, 98, 99, 100, 101, 102, 103, 104, 105, 106, 107, 109, 110, 111, 112, 113, 114, 115, 116, 117, 118, 119, 120, 121, 122, 123, 124, 125, 126, 127, 128, 129, 131, 132, 143, 145, 146, 151, 152, 153, 154, 155, 156, 157, 159, 160, 162, 163, 164, 165, 166
English as a second language 2, 142, 152, 153, 157, 158
English as an additional language 2, 3, 4, 7, 16, 19, 20, 30, 39, 46, 67
English language learners 2, 145, 146, 157, 158
English National Curriculum 23, 24
English-medium schools 17, 47-62, 114, 117
ESL (*see* English as a second language)
ESL curriculum 158
ethnography 7, 45, 123, 126
Evans, S. 125
Evans, Z. 123
everyday language 10, 19, 27, 35, 52, 58, 60-61, 109-110, 125, 137, 145, 148, 151
explanations 7, 8, 32, 49, 80, 92, 94, 96, 108, 121, 125, 129, 136, 137, 141, 156, 159, 162
exponent 35, 44

fair share 56, 59, 60
fair testing 125-126
Faltis, C. 118, 120
fansaar 45
Farrugia, M. 11, 98, 115, 163, 164
Farsi 25
Favilli, F. 44
Finnish 38
Franceschini, R. 15
Frederickson, N. 123
function 44, 45, 166

Garcia, E. 132
Gee, J. 78
geeso 35
gender 33, 37
General Certificate of Education (A-levels) 125
General Certificate of Secondary Education (GCSE) 123-125
genre 23, 66, 69, 71, 129, 124, 167
Gerofsky, S. 66, 72
gestures 19, 35, 44, 78, 82, 83, 85, 86, 87, 88, 95, 108, 109, 121
Glaser, B. 135
globalisation 48

Gorgorió, N. 50
grammar 57, 58, 59
grammatical errors 110, 140
graphs 11, 34, 78-96, 105, 108, 109, 162, 167
Greek 35, 145
Griffiths, R. 107
Grosjean, F. 80
Gujarati 16
Gujrati 51
Gumperz, J. 121
Gutstein, E. 32

Haan Associates 39
Halai, A. 10, 11, 27, 48, 51, 100, 151, 162, 163, 164
Halcón, L. 36, 37
Halliday, M. 10, 14, 15, 19, 20, 27, 109, 146
Hambleton, R. 124
Haque, A. 47
Hargreaves, E. 3
Hart, K. 56
Harvey, R. 109
Hasan Aly, J. 48
Heath, S. 45
Heller, M. 120, 123
hidden curriculum 20, 50, 60
Hindu-Arabic numerals 28
Hmong 32
home language(s) 5, 6, 12, 13, 17, 18, 32, 44, 46, 99, 137, 152, 154, 161, 164
Hornberger, N. 165
Howie, S. 3, 4
Howson, A. 146
Hughes, T. 123
Hunston, S. 21
Hutchins, P. 159

idiom 147, 153, 155
if and only if 109
if…then 5, 134, 156
immersion education 4, 79, 102, 114
imperative mood 109
inclusion 11, 17, 97, 106-107, 111
inequality (mathematical) 33, 40
Initiation-Response-Feedback 103-104
interactional sociolinguistics 121
International Test Commission 124
Iowa Test of Basic Skills 143
isgoys 34
iskudofa 44
Italian 5, 34, 40, 101, 152, 157

Jacobson, R. 120
Jama Musse, J. 34, 44
Jamaican Creole 5
jibbaar 35, 44
Johnson, J. 34
Jones, D. 11, 19, 111, 114, 119, 120, 121, 124, 162, 163, 166

Jones, S. 120

Karplus, R. 55
Katchi 51
Keenan, P. 101
Khan, A. 47
Khan, M. 47
Khisty, L. 7, 12, 133, 144, 161, 162, 164, 167
Kress, G. 21

L1 (*see* home language)
Lakoff, G. 34, 45
lammaane 34
Lamon, S. 94
language
– acquisition 28, 45, 131-133, 140, 143
– and mathematics (*see* mathematics and language)
– as a resource 44, 62
– dominant 4, 8, 50, 61, 62, 80, 128, 145, 146, 164
– informal 6, 44, 109, 110, 147, 148-150, 155, 157, 161
– functions 109, 144
– learning 4, 11, 12, 14, 37, 38, 61, 63, 64, 73, 74, 75, 77, 79, 98, 133, 135, 140, 143, 144, 145, 146, 167
– of examinations 3, 4, 100, 123-126
– of instruction 32, 34, 37-62, 97-111, 114, 154
– of textbooks 3, 51, 100
– official 3, 6, 11, 47, 50, 97, 111, 113, 153, 161, 163
– policy 50, 106, 112, 153, 154, 157
– politics 9, 10, 16, 46, 47-62, 101, 113, 126
language proficiency 3-6, 12, 28, 37, 47, 48, 49, 57, 58, 59, 61, 62, 63, 64, 65, 67, 79, 80, 106, 115, 116, 122, 129, 145, 146, 154, 155, 157, 163
– and mathematical attainment 3-6
– English 3-6, 28, 37, 47, 48, 59, 63, 64, 65, 67, 79, 80, 106, 115, 129, 146, 157
– of teachers 58, 61, 62, 150
– Spanish 80, 129
– Welsh 12, 116, 122, 162, 164
Lapkin, S. 74
Laporte, R. 47
Lappan, G. 81, 82
Latin 35
Latino/as 12, 32, 80, 128-144, 177
Lave, J. 64
LeBlanc, K. 38
Lee, C. 109
Leung, C. 44, 45
lexical syllabus 30
Lilburn, P. 156
Lindholm-Leary, K. 120
line (mathematical) 23, 26, 35, 109
linguistic bias 103

linguistic interviews 33, 40, 42
literacy 3, 25, 38
– mathematical 26
– multi-modal 25
Lodholz, R. 133
logical connectives 5, 30, 156
logical thinking 5, 48, 49, 61

Makoni, S. 165
Malta 11, 97-112, 163
Maltese 11, 97-112, 163, 164
Maltese National Minimum Curriculum 97-112
Maōri 34, 111
mapping (mathematical) 20, 26
Marks, G. 134
Martin, J. 27
Martin-Jones, M. 120
matched assessments 124-125
mathematical concepts 25, 34, 43, 44, 56, 60, 61, 79, 93, 95, 167
mathematical conventions 20, 25, 135, 136, 142-143
mathematical discussion 4, 7, 46, 78, 79-81, 93, 95, 96, 112, 137, 154, 164
mathematical English 10, 20, 25, 104, 107, 111, 155, 165
mathematical justification 49, 156
mathematical langscape 14-31
mathematical meaning 11, 12, 14, 24, 25, 27, 30, 44, 45, 56, 59, 80, 134, 149, 162, 166
mathematical practices 4, 45, 144
mathematical reasoning (*see* reasoning)
mathematical register 14, 27, 109-111, 125, 155, 166-167
– Welsh 125
– written 109-111
mathematical terminology (*see* terminology)
mathematical texts 22-23, 52, 58, 60
mathematical thinking 5, 6, 11, 12, 34, 43, 44, 48, 49, 59, 61, 64, 69, 73, 74, 75, 134, 137, 141, 147, 155, 156, 165
mathematical understanding 9, 10, 11, 33, 35, 39, 44, 52, 56, 58, 59, 80, 93, 101, 119, 121, 135, 143, 151, 161
mathematical word problems 5, 11, 63-77, 100, 134, 167
mathematicians 166
mathematics
– and culture 10, 33, 39, 43-45, 159, 165
– and language 6, 20, 23, 30, 135, 140, 146, 161, 165, 167
– assessment 12, 66, 123-126
– attainment 1, 2, 3-6, 65, 117, 124-127
– curriculum 11, 14, 23, 20, 31, 48, 61, 68, 153
– nature of 165
– teacher educators 61
– test development 123-126

mathematics teaching
– classroom management 121, 122
– dilemmas 7-8, 49
– formal 8, 9, 36, 48, 51, 61
– grouping 116-118
– materials and resources 14, 22, 23, 24, 27-30, 31, 46, 51, 61, 119-120
– use of literature 159
– lesson structure 118-119
matrix 35
meaning, mathematical (*see* mathematical meaning)
meaning, multiple 45, 78, 92, 93, 95
measures 35, 42, 44, 55, 94, 104, 108
medium of instruction 47, 48, 50, 61, 98-111, 114
Mehan, H. 95
Mejia, A. 120
Mendes, J. 64
Mercer, N. 16
Mestre, J. 64, 65, 77, 146
metaphor 10, 34-35, 40, 42-45, 109, 166
migration 2, 32, 34, 37, 46
Miller, L. 109
Minglish 47, 60
Ministry of Education (Malta) 97, 99, 101, 103, 104, 106, 111, 112
Ministry of Education (Pakistan) 48, 61
Ministry of Education and National Culture (Malta) 112
Ministry of Education, Youth and Employment (Malta) 112
minority rights 2
Mirpuri 5
Mohan, B. 133, 134, 143
Moll, L. 136, 144
Monaghan, F. 10, 15, 16, 23, 31, 109, 134, 164, 166, 167
monolingual learners 3, 6, 65, 69,
monolingual teachers 15, 96, 150, 153, 157
Morgan, C. 15, 32, 66, 109, 110, 134, 166
Moschkovich, J. 6, 7, 11, 32, 44, 45, 80, 115, 144, 146, 162, 165
Mousley, J. 134
multilingual teachers 150
multilingualism 1-4, 6, 12, 126, 165, 167
Multiliteracy Project, The 168, 170
multimodal texts 21, 25
multiple interpretation 78-96
multiple language repertoires 165-166
multiply 20, 44, 137

National Center for Educational Statistics (NCES) (US) 128
National Council of Teachers of Mathematics (NCTM) (US) 48, 61
National Curriculum Tests (Wales) 124-125, 127
native language 97, 113, 114, 115

Index

native speakers 26, 78, 79, 115, 123
Nepali 17-18
Noelting, G. 56
nominalisation 94, 109
Núñez, R. 34, 35, 45

O'Halloran, K. 21
Oakland, T. 124
ongl sgwâr 125
ordinary meaning 24, 25, 35, 43, 44, 45

Pakistan 1, 8, 10, 16, 47-62, 67, 74, 151, 163
Pakistan National Curriculum 61
Papua New Guinea 5, 13, 151, 159
parabola 44
Parvanehnezhad, Z. 146, 152
passive voice 109
pedrochr 124
Peregoy, S. 133, 134
perimeter 108, 109, 129, 127
Phillips, C. 3
pie charts 29-30
Pike, N. 123
Pimm, D. 10,15, 44, 107, 109, 134, 146, 147
plan view 124
Planas, N. 50
polyhedron 35
polysemy 40, 44
positioning 50
Potter, J. 68, 70
power 48, 50, 61
procedural discourse 8, 121, 129, 134
pronunciation 73, 74, 138
proof 30
proportion 56, 59, 60
Pujolar, J. 123
punctuation 73, 129, 142
Punjabi 1, 4, 5, 6, 8

qaanso 34
QCA (Qualifications and Curriculum Authority) (UK) 63
quadrilateral 124
quantitative reasoning (*see* reasoning, quantitative)
Queensland Studies Authority (Australia) 148
Quinn, R. 134

Rahman, T. 47, 50
ratio 19, 52-56, 59
reading comprehension 5
real world thinking 56, 64, 66, 69, 74
reasoning
– additive 56
– deductive 5
– mathematical 5, 60
– multiplicative 55, 56
– quantitative 56

rectangle 129, 137, 138
Reimers, F. 48
Renouf, A. 30
rhetorical strategies 70
right angles 125
right triangle 129, 127-128
Roberts, G. 123
Roberts, T. 111
Robillos, M. 37, 38, 43
Rosebery, A. 79
rotation 25-27, 109
Rothman, R. 107

saab 44
Sams, C. 16
scales (mathematical) 19, 78-96, 162, 165
Schwartzman, S. 35
Secada, W. 4, 64, 65, 146
second language acquisition 28, 131, 134
second language learners 116, 128, 129, 132, 133, 134
sense making 64, 69, 71, 75
Setati, M. 4, 8, 9, 44, 49, 50, 51, 59, 102, 104, 126, 151
Setswana 8, 9, 102
Sfard, A. 35, 56
Sierpinska, A. 49
Silver, E. 134
similar 19, 20
Sinclair, J. 30, 103
Sindhi 47, 48, 51
Slater, T. 133, 134, 143
slope 35, 94, 109
Slovakian 17
SMILE mathematics scheme 23-24, 26, 27, 28, 30, 31
social elite 61
Social Policy Development Centre (SPDC) (Pakistan) 48
sociolinguistic contexts 1-2, 8, 12
solidarity 8
Somali 10, 32-46, 166-167
Somalia 32-46, 67, 145
– educational history 34-38
South Africa 2, 3, 7, 8, 50, 102, 104, 115
Sowder, J. 56
spell check 143
spelling 18, 73, 74, 129, 140, 142
Stodolsky, S. 95
Strauss, A. 135
Street, B. 25, 26
Sullivan, P. 156
Swain, M. 74
symbols (mathematical) 14, 25, 34, 35, 40-42, 134, 148, 156, 159
syntax 19, 20, 27, 95, 134

teacher assessment 123, 127, 135
teaching dilemmas 7, 8, 38, 43, 49, 97, 105

technology (in education) 37, 38, 97, 120
technology (school subject) 97, 101
tense 10, 57, 59, 73, 129, 163
terminology 33-39, 44, 98, 124, 125, 134 (*see also* vocabulary)
Thomas, J. 159
Thomas, W. 64
Thompson, A. 95
threshold hypothesis 4-5
through 26
time, notions of 159
TIMSS-R 3
Tok Pisin 151
Tollefson, J. 97, 101
translation (linguistic) 20, 33, 39, 53, 54, 56, 58, 100, 101, 110, 118, 120, 124, 125, 127, 151, 159
transparency 7
Tsui, A. 97, 101

UNESCO 14
United Kingdom 2, 3, 4, 5, 11, 12, 14, 15, 16, 17, 19, 25, 113, 123
United States 7, 10, 11, 12, 32, 36, 37, 42, 45, 65, 79, 80, 115, 128, 132, 133, 164
unitizing 93-95
Urdu 1, 6, 8, 10, 16, 47, 48, 51-54, 59-62, 151, 162, 163, 164
uwcholwg 124

van Leeuwen, T. 21
Verschaffel, L. 64
vertex 23, 35
Victorian Curriculum and Assessment Authority (Australia) 148
Vietnamese 145, 151, 152

vocabulary 7, 10, 19, 22, 23-25, 30, 33-35, 36, 44-46, 58-59, 65, 77, 99, 101, 144
– mathematical 7, 10, 33-35, 44-46, 98, 109-111
Vygotsky, L. 129, 134, 144

Wales 3, 11, 67, 113-127, 163
Warren, B. 79
Warwick, D. 48
Waters, T. 38
Welsh 11, 12, 111, 113-127, 162, 163, 164, 166
Welsh Assembly Government 114
Welsh Language Act 1993 113
Welsh Language Board 113
Welsh-medium schools 12, 113, 114-119, 127 (*see also Ysgolion Cymraeg*)
Whagi 151
will 57-59
Williams, C. 118, 119, 120
Willis, D. 30
Wilson, M. 134
word problems (*see* mathematical word problems)
writing 12, 18, 25, 26, 27, 30, 35, 44, 67-77, 105, 107, 121, 122, 128-144, 147, 158, 161, 162, 168

xariiq 35
xijaab 33, 37

Young, B. 128
Ysgolion Cymraeg 114

Zaskis, R. 107
ziada 53-55, 59